Generalized Vector and Dyadic Analysis

Applied Mathematics in Field Theory

Chen-To Tai

Radiation Laboratory
Department of Electrical Engineering
and Computer Science
University of Michigan

IEEE Antennas and Propagation Society, *Sponsor*

IEEE
PRESS

The Institute of Electrical and Electronics Engineers, Inc., New York

IEEE PRESS
445 Hoes Lane, PO Box 1331
Piscataway, NJ 08855-1331

ISBN 0-87942-288-2
IEEE Order Number: PC0283-2

Printed in the United States of America

10 9 8 7 6 5 4 3 2

Library of Congress Cataloging-in-Publication Data

Tai, Chen-To (date)
Generalized vector and dyadic analysis / Chen-To Tai.
p. cm.
"IEEE order number: PC0283-2"—T.p. verso.
Includes bibliographical references and index.
ISBN 0-87942-288-2
1. Vector analysis. I. Title. II. Title: Dyadic analysis.
QA433.T3 1992
515'.63—dc20 91-23537

Contents

In Memory
of
Professor Dr. Yeh Chi-Sun
(1898–1977)

Preface

Mathematics is a language.

The whole is simpler than its parts.

Anyone having these desires will make these researches.

—J. Willard Gibbs

This monograph is mainly based on the author's recent work on vector analysis and dyadic analysis. The book is divided into two main topics: Chapters 1–6 cover vector analysis, while Chapter 7 is exclusively devoted to dyadic analysis. On the subject of vector analysis, a new symbolic method with the aid of a symbolic vector is the main feature of the presentation. By means of this method, the principal topics in vector analysis can be developed in a systematic way. All vector identities can be derived by an algebraic manipulation of expressions with two partial symbolic vectors without actually performing any differentiation. Integral theorems are formulated under one roof with the aid of a generalized Gauss theorem. Vector analysis on a surface is treated in a similar manner. Some basic differential functions on a surface are defined; they are different from the surface functions previously defined by Weatherburn, although the two sets are intimately related. Their relations are discussed in great detail. The advantage of adopting the surface functions advocated in this work is the simplicity of formulating the surface integral theorems based on these newly defined functions.

The scope of topics covered in this book on vector analysis is comparable to those found in the books by Wilson [21], Gans [4], and Phillips [11]. However, the topics on curvilinear orthogonal systems have been treated in great detail. One important feature of this work is the unified treatment of many theorems and formulas of similar nature, which includes the invariance principle of the differential operators for the gradient, the divergence, and the curl, and the relations between various integral theorems and transport theorems. Some quite useful topics are found in this book, which include the derivation of several identities involving the derivatives of unit vectors, and the relations between the unit vectors of various coordinate systems based on a method of gradient.

Tensor analysis is outside the scope of this book. There are many excellent books treating this subject. Since dyadic analysis is now used quite frequently in engineering sciences, a chapter on this subject, which is closely related to tensor analysis in a three-dimensional Euclidean space, may be timely.

As a whole, it is hoped that this book may be useful to instructors and students in engineering and physical sciences who wish to teach and to learn vector analysis in a systematic manner based on a new method with a clear picture of the constituent structure of this mature science not critically studied in the past few decades.

Acknowledgment

Without the encouragement which I received from my wife and family, and the loving innocent interference from my grandchildren, this work would never have been completed. I would like to express my gratitude to President Dr. Qian Wei-Chang for his kindness in inviting me as a Visiting Professor at The Shanghai University of Technology in the Fall of 1988 when this work was started. Most of the writing was done when I was a Visiting Professor at The Chung Cheng Institute of Technology, Taiwan, in the Spring of 1990. I am indebted to President Dr. Chen Chwan-Haw, Prof. Bor Sheau-Shong, and Prof. Kuei Ching-Ping for the invitation.

The assistance of Prof. Nenghang Fang of The Nanjing Institute of Electronic Technology, China, currently a Visiting Scholar at The University of Michigan, has been most valuable. His discussion with me about the Russian work on vector analysis was instrumental in stimulating my interest to formulate the symbolic vector method introduced in this book. Without his participation in the early stage of this work, the endeavor could not have begun. He has kindly checked all the formulas and made numerous suggestions. I am grateful to many colleagues for useful information and valuable comments. They include: Prof. J. Van Bladel of The University of Gent, Prof. Jed Z. Buchwald of The University of Toronto, Prof. W. Jack Cunningham of Yale University, Prof. Walter R. Debler and Prof. James F. Driscoll of The University of Michigan, Prof. John D. Kraus and Prof. H. C. Ko of The Ohio State University, and Prof. C. Truesdell of The Johns Hopkins University. My dear old friend Prof. David K. Cheng of Syracuse University kindly edited the manuscript and suggested the title of the book. The teachings of Prof. Chih-Kung Jen of The Johns Hopkins Applied

Physics Laboratory, formerly of Tsing Hua University, and Prof. Ronold W. P. King of Harvard University remain the guiding lights in my search for knowledge. Without the help of Ms. Bonnie Kidd, Dr. Jian-Ming Jin, and Dr. Leland Pierce, the preparation of this manuscript would not have been so professional and successful.

I wish to thank Prof. Fawwaz T. Ulaby, Director of the Radiation Laboratory, for providing me with technical assistance. The speedy production of this book is due to the efficient management of Mr. Dudley Kay, Executive Editor, and the valuable technical supervision of Ms. Anne Reifsnyder, Associate Editor, of the IEEE Press. Some major changes have been made in the original manuscript as a result of many valuable suggestions from the reviewers. I am most grateful to these reviewers.

CHEN-TO TAI

Ann Arbor, MI

Publisher's Acknowledgment

The IEEE Press would like to thank Dr. Wilson Pearson of Clemson University, Antennas and Propagation Society Liaison to the Press, for his support of the work. We would also like to thank Robert D. Nevels of Texas A & M University, Glenn S. Smith of Georgia Institute of Technology, and N. G. Alexopoulos of the University of California, Los Angeles for their constructive comments as reviewers of the manuscript.

CHAPTER 1

Vector Algebra

1-1 REPRESENTATIONS OF VECTOR FUNCTIONS

A vector function has both magnitude and direction. The vector functions which we encounter in many physical problems are, in general, functions of space and time. In the first five chapters, we discuss only their characteristics as functions of spatial variables. Functions of space and time are covered in Chapter 6 dealing with a moving surface or a moving contour.

A vector function is denoted by $\mathbf{F}$. Geometrically, it is represented by a line with an arrow in a three-dimensional space. The length of the line corresponds to its magnitude, and the direction of the line represents the direction of the vector function. The convenience of using vectors to represent physical quantities is illustrated by a simple example shown in Fig. 1-1 which describes the motion of a mass particle in a frictionless air (vacuum) against a constant gravitational force. The particle is thrown into the space with an initial velocity $\mathbf{v}_0$, making an angle θ_0 with respect to the horizon. During its flight, the velocity function of the particle changes both its magnitude and direction, as shown by $\mathbf{v}_1, \mathbf{v}_2$, etc., at subsequent locations. The gravitational force which acts on the particle is assumed to be constant, and it is represented by $\mathbf{F}$ in the figure. A constant vector function means that both the magnitude and the direction of the function are constant, being independent of the spatial variables, x and z in this case.

The rule of the addition of two vectors $\mathbf{a}$ and $\mathbf{b}$ is shown geometrically by Fig. 1-2 (a), (b), or (c). Algebraically, it is written in the same form as the addition of two numbers of two scalar functions, i.e.,

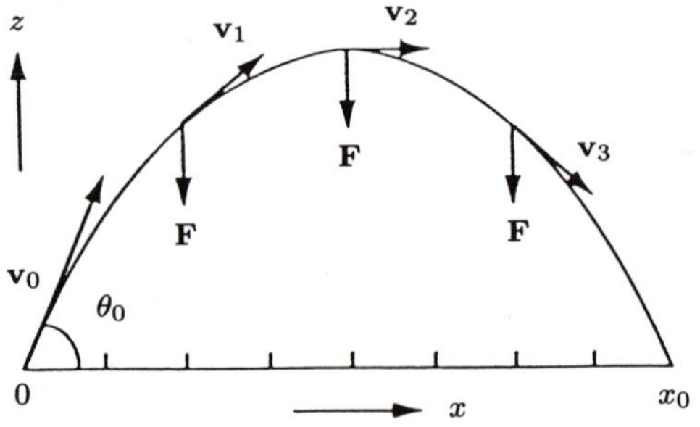

$$z = x\left(1 - \frac{x}{x_0}\right)\tan\theta_0$$

$$x_0 = \frac{1}{g}v_0^2 \sin 2\theta_0$$

$$g = \text{gravitational constant}$$

Fig. 1-1 Trajectory of a mass particle in a gravitational field showing the velocity $\mathbf{v}$ and the constant force vector $\mathbf{F}$ at different locations.

$$\mathbf{c} = \mathbf{a} + \mathbf{b}. \tag{1.1}$$

The subtraction of vector **b** from vector **a** is written in the form

$$\mathbf{d} = \mathbf{a} - \mathbf{b}. \tag{1.2}$$

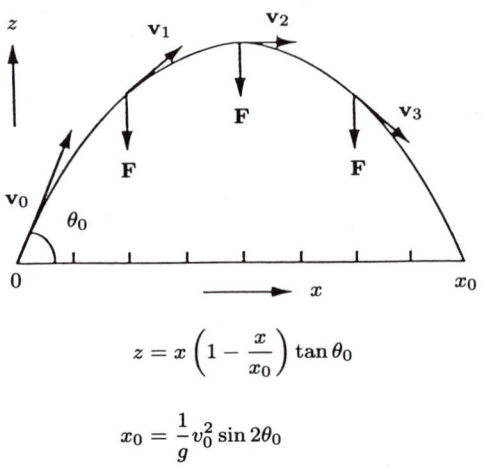

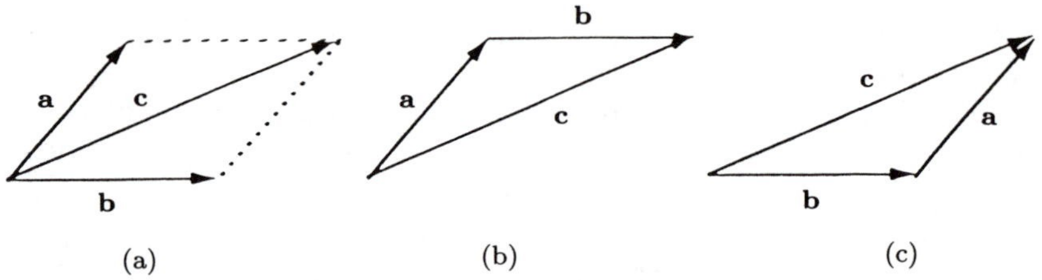

Fig. 1-2 Addition of vectors, $\mathbf{a} + \mathbf{b} = \mathbf{c}$.

Now, $-\mathbf{b}$ is a vector which has the same magnitude as **b**, but of opposite direction; then (1.2) can be considered as the addition of **a** and $(-\mathbf{b})$. Geometrically, the meaning of (1.2) is shown in Fig. 1-3. The sum and the difference of two vectors obey the associate rule, i.e.,

$$\mathbf{a} + \mathbf{b} = \mathbf{b} + \mathbf{a} \tag{1.3}$$

and

$$\mathbf{a} - \mathbf{b} = -\mathbf{b} + \mathbf{a}. \tag{1.4}$$

They can be generalized to any number of vectors.

The rule of the addition of vectors suggests that any vector can be considered as being made of basic components associated with a proper coordinate

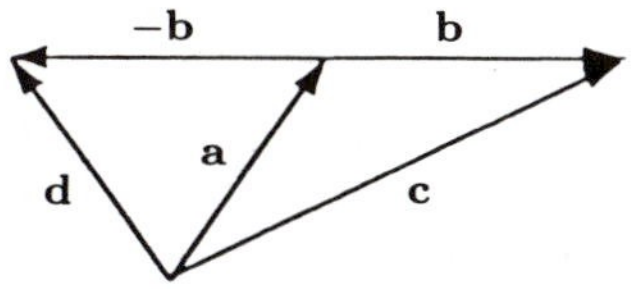

Fig. 1-3 Subtraction of vectors, $\mathbf{a} - \mathbf{b} = \mathbf{d}$.

system. The most convenient system to use is the Cartesian system or the rectangular coordinate system. The spatial variables in this system are commonly denoted by x, y, z. A vector which has a magnitude equal to unity and pointed in the positive x direction is called a unit vector in the x direction and is denoted by $\mathbf{u}_x$. Similarly, we have $\mathbf{u}_y, \mathbf{u}_z$. In such a system, a vector function $\mathbf{F}$ which, in general, is a function of position, can be written in the form

$$\mathbf{F} = F_x \mathbf{u}_x + F_y \mathbf{u}_y + F_z \mathbf{u}_z. \tag{1.5}$$

The three scalar functions F_x, F_y, F_z are called the components of $\mathbf{F}$ in the direction of $\mathbf{u}_x, \mathbf{u}_y$, and $\mathbf{u}_z$, respectively, while $F_x\mathbf{u}_x$, $F_y\mathbf{u}_y$, and $F_z\mathbf{u}_z$ are called the vector components of $\mathbf{F}$. The geometrical representation of $\mathbf{F}$ is shown in Fig. 1-4. It is seen that F_x, F_y, and F_z can be either positive or negative. In Fig. 1-4, F_x and F_z are positive, but F_y is negative.

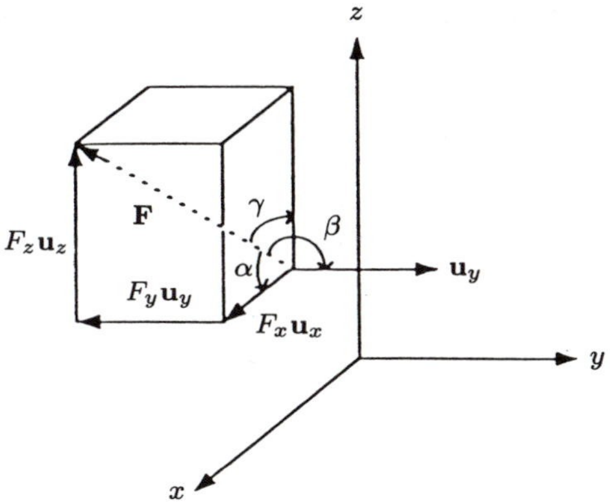

Fig. 1-4 Components of a vector Cartesian system.

In addition to the representation by (1.5), it is sometimes desirable to express $\mathbf{F}$ in terms of its magnitude, denoted by $|\mathbf{F}|$, and its directional cosines, i.e.,

$$\mathbf{F} = |\mathbf{F}| \left(\cos\alpha \mathbf{u}_x + \cos\beta \mathbf{u}_y + \cos\gamma \mathbf{u}_z\right). \tag{1.6}$$

zα, β, and γ are the angles which $\mathbf{F}$ makes, respectively, with $\mathbf{u}_x, \mathbf{u}_y$, and $\mathbf{u}_z$, as shown in Fig. 1-4. It is obvious from the geometry of that figure that

$$|\mathbf{F}| = \left(F_x^2 + F_y^2 + F_z^2\right)^{\frac{1}{2}} \tag{1.7}$$

and

$$\cos\alpha = \frac{F_x}{|\mathbf{F}|}, \cos\beta = \frac{F_y}{|\mathbf{F}|}, \cos\gamma = \frac{F_z}{|\mathbf{F}|}. \tag{1.8}$$

Furthermore, we have the relation

$$\cos^2\alpha + \cos^2\beta + \cos^2\gamma = 1. \tag{1.9}$$

In view of (1.9), only two of the directional cosine angles are independent. From the above discussion, we observe that, in general, we need three parameters to specify a vector function. The three parameters could be F_x, F_y, and F_z or $|\mathbf{F}|$ and two of the directional cosine angles. Representations such as (1.5) and (1.6) can be extended to other orthogonal coordinate systems which will be discussed in a later chapter.

1-2 PRODUCTS AND IDENTITIES

The scalar product of two vectors **a** and **b** is denoted by $\mathbf{a}\cdot\mathbf{b}$ and it is defined by

$$\mathbf{a}\cdot\mathbf{b} = |\mathbf{a}|\,|\mathbf{b}|\cos\theta \tag{1.10}$$

where θ is the angle between **a** and **b**, as shown in Fig. 1-5. Because of the notation used for such a product, sometimes it is called the dot product. By applying (1.10) to three orthogonal unit vectors $\mathbf{u}_1, \mathbf{u}_2, \mathbf{u}_3$, one finds

$$\mathbf{u}_i\cdot\mathbf{u}_j = \left\{\begin{array}{ll} 1, & i=j \\ 0, & i\neq j \end{array}\right\}, \qquad i,j = 1,2,3. \tag{1.11}$$

The value of $\mathbf{a}\cdot\mathbf{b}$ can also be expressed in terms of the components of **a** and **b** in any orthogonal system. Let the system under consideration be the Cartesian system, and let $\mathbf{c} = \mathbf{a} - \mathbf{b}$; then

$$|\mathbf{c}|^2 = |\mathbf{a}-\mathbf{b}|^2 = |\mathbf{a}|^2 + |\mathbf{b}|^2 - 2\,|\mathbf{a}|\,|\mathbf{b}|\cos\theta.$$

Hence,

$$\begin{aligned} \mathbf{a}\cdot\mathbf{b} &= |\mathbf{a}|\,|\mathbf{b}|\cos\theta = \frac{|\mathbf{a}|^2 + |\mathbf{b}|^2 - |\mathbf{a}-\mathbf{b}|^2}{2} \\ &= \frac{a_x^2 + a_y^2 + a_z^2 + b_x^2 + b_y^2 + b_z^2 - (a_x-b_x)^2 - (a_y-b_y)^2 - (a_z-b_z)^2}{2} \\ &= a_xb_x + a_yb_y + a_zb_z. \end{aligned} \tag{1.12}$$

By equating (1.10) and (1.12), one finds

$$\begin{aligned} \cos\theta &= \frac{1}{|\mathbf{a}|\,|\mathbf{b}|}(a_xb_x + a_yb_y + a_zb_z) \\ &= \cos\alpha_a\cos\alpha_b + \cos\beta_a\cos\beta_b + \cos\gamma_a\cos\gamma_b, \end{aligned} \tag{1.13}$$

a relationship well known in analytical geometry. Equation (1.12) can be used to prove the validity of the distributive law for the scalar products, namely,

$$(\mathbf{a}+\mathbf{b})\cdot\mathbf{c} = \mathbf{a}\cdot\mathbf{c}+\mathbf{b}\cdot\mathbf{c}. \tag{1.14}$$

According to (1.12), we have

$$\begin{aligned}(\mathbf{a}+\mathbf{b})\cdot\mathbf{c} &= (a_x+b_x)\,c_x+(a_y+b_y)\,c_y+(a_z+b_z)\,c_z\\ &= (a_xc_x+a_yc_y+a_zc_z)+(b_xc_x+b_yc_y+b_zc_z)\\ &= \mathbf{a}\cdot\mathbf{c}+\mathbf{b}\cdot\mathbf{c}.\end{aligned}$$

Once we have proved the distributive law for the scalar product, (1.12) can be verified by taking the sum of the scalar products of the individual terms of **a** and **b**.

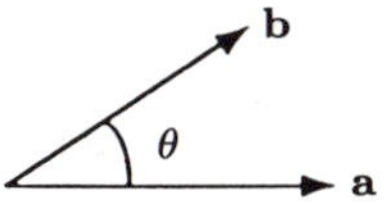

Fig. 1-5 Scalar product of two vectors, $\mathbf{a}\cdot\mathbf{b} = |\mathbf{a}||\mathbf{b}|\cos\theta$.

The vector product of two vector functions **a** and **b**, denoted by $\mathbf{a}\times\mathbf{b}$, is defined by

$$\mathbf{a}\times\mathbf{b} = |\mathbf{a}|\,|\mathbf{b}|\sin\theta\mathbf{u}_c \tag{1.15}$$

where θ denotes the angle between **a** and **b**, measured from **a** to **b**; $\mathbf{u}_c$ denotes a unit vector perpendicular to both **a** and **b** and is pointed to the advancing direction of a right-hand screw when we turn from **a** to **b**. Figure 1-6 shows the relative position of $\mathbf{u}_c$ with respect to **a** and **b**. Because of the notation used for the vector product, it is sometimes called the cross product, in contrast to the dot product or the scalar product. For three orthogonal unit vectors in a right-hand system, we have $\mathbf{u}_1\times\mathbf{u}_2 = \mathbf{u}_3, \mathbf{u}_2\times\mathbf{u}_3 = \mathbf{u}_1$, and $\mathbf{u}_3\times\mathbf{u}_1 = \mathbf{u}_2$. It is obvious that $\mathbf{u}_i\times\mathbf{u}_i = 0, i = 1,2,3$. From the definition of the vector product defined by (1.15), one finds

$$\mathbf{b}\times\mathbf{a} = -\mathbf{a}\times\mathbf{b}. \tag{1.16}$$

The value of $\mathbf{a}\times\mathbf{b}$ as described by (1.15) can also be expressed in terms of the components of **a** and **b** in a Cartesian coordinate system. If we let $\mathbf{a}\times\mathbf{b} = \mathbf{v} = v_x\mathbf{u}_x+v_y\mathbf{u}_y+v_z\mathbf{u}_z$, which is perpendicular to both **a** and **b**, then

$$\mathbf{a}\cdot\mathbf{v} = a_xv_x+a_yv_y+a_zv_z = 0 \tag{1.17}$$

$$\mathbf{b}\cdot\mathbf{v} = b_xv_x+b_yv_y+b_zv_z = 0. \tag{1.18}$$

Solving for v_x/v_z and v_y/v_z, from (1.17) and (1.18) we obtain

$$\frac{v_x}{v_z} = \frac{a_yb_z-a_zb_y}{a_xb_y-a_yb_x},\quad \frac{v_y}{v_z} = \frac{a_zb_x-a_xb_z}{a_xb_y-a_yb_x}.$$

Thus,

$$\frac{v_x}{a_yb_z-a_zb_y} = \frac{v_y}{a_zb_x-a_xb_z} = \frac{v_z}{a_xb_y-a_yb_x}.$$

Let the common ratio of these quantities be denoted by c, which can be determined by considering the case with $\mathbf{a} = \mathbf{u}_x, \mathbf{b} = \mathbf{u}_y$; then $\mathbf{v} = \mathbf{a} \times \mathbf{b} = \mathbf{u}_z$; hence, from the last ratio, we find $c = 1$ because $v_z = 1$ and $a_x = b_y = 1$ while $a_y = b_x = 0$. The three components of $\mathbf{v}$, therefore, are given by

$$\left.\begin{aligned} v_x &= a_y b_z - a_z b_y \\ v_y &= a_z b_x - a_x b_z \\ v_z &= a_x b_y - a_y b_x \end{aligned}\right\} \tag{1.19}$$

which can be assembled in a determinant form as

$$\mathbf{v} = \begin{vmatrix} \mathbf{u}_x & \mathbf{u}_y & \mathbf{u}_z \\ a_x & a_y & a_z \\ b_x & b_y & b_z \end{vmatrix}. \tag{1.20}$$

We can use (1.20) to prove the distributive law of vector products, i.e.,

$$(\mathbf{a} + \mathbf{b}) \times \mathbf{c} = \mathbf{a} \times \mathbf{c} + \mathbf{b} \times \mathbf{c}. \tag{1.21}$$

To prove (1.21), we find that the x component of $(\mathbf{a} + \mathbf{b}) \times \mathbf{c}$ according to (1.20) is equal to

$$\begin{aligned} &(a_y + b_y)\, c_z \;-\; (a_z + b_z)\, c_y \\ &= (a_y c_z - a_z c_y) \;+\; (b_y c_z - b_z c_y)\,. \end{aligned} \tag{1.22}$$

The last two terms in (1.22) denote, respectively, the x component of $\mathbf{a} \times \mathbf{c}$ and $\mathbf{b} \times \mathbf{c}$. The equality of the y and z components of (1.21) can be proved in a similar manner.

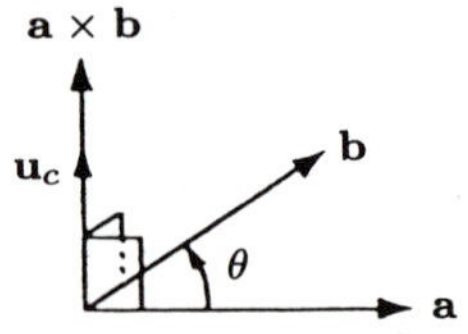

Fig. 1-6 Vector product of two vectors, $\mathbf{a} \times \mathbf{b} = |\mathbf{a}||\mathbf{b}| \sin\theta \mathbf{u_C}$; $\mathbf{u_C} \perp \mathbf{a}$, $\mathbf{u_C} \perp \mathbf{b}$.

In addition to the scalar product and the vector product introduced before, there are two identities involving the triple products that are very useful in vector analysis. They are

$$\mathbf{a} \cdot (\mathbf{b} \times \mathbf{c}) = \mathbf{b} \cdot (\mathbf{c} \times \mathbf{a}) = \mathbf{c} \cdot (\mathbf{a} \times \mathbf{b}) \tag{1.23}$$

$$\mathbf{a} \times (\mathbf{b} \times \mathbf{c}) = (\mathbf{a} \cdot \mathbf{c})\, \mathbf{b} - (\mathbf{a} \cdot \mathbf{b})\, \mathbf{c}. \tag{1.24}$$

Identities described by (1.23) can be proved by writing $\mathbf{a} \cdot (\mathbf{b} \times \mathbf{c})$ in a determinant form:

$$\mathbf{a} \cdot (\mathbf{b} \times \mathbf{c}) = \begin{vmatrix} a_1 & a_2 & a_3 \\ b_1 & b_2 & b_3 \\ c_1 & c_2 & c_3 \end{vmatrix}.$$

According to the theory of determinants,

$$\begin{vmatrix} a_1 & a_2 & a_3 \\ b_1 & b_2 & b_3 \\ c_1 & c_2 & c_3 \end{vmatrix} = \begin{vmatrix} b_1 & b_2 & b_3 \\ c_1 & c_2 & c_3 \\ a_1 & a_2 & a_3 \end{vmatrix} = \begin{vmatrix} c_1 & c_2 & c_3 \\ a_1 & a_2 & a_3 \\ b_1 & b_2 & b_3 \end{vmatrix}.$$

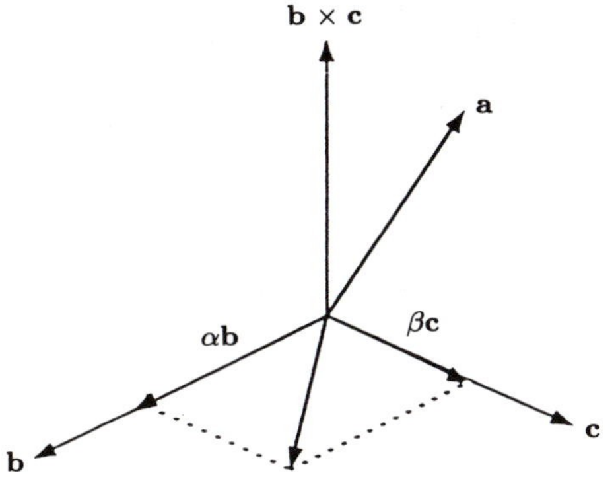

Fig. 1-7 Orientation of various vectors in $\mathbf{a} \times (\mathbf{b} \times \mathbf{c})$.

The last two determinants represent, respectively, $\mathbf{b} \cdot (\mathbf{c} \times \mathbf{a})$ and $\mathbf{c} \cdot (\mathbf{a} \times \mathbf{b})$; hence, we have the validity of (1.23). To prove (1.24), we observe that the vector $\mathbf{a} \times (\mathbf{b} \times \mathbf{c})$ lies in the plane containing $\mathbf{b}$ and $\mathbf{c}$, so we can treat $\mathbf{a} \times (\mathbf{b} \times \mathbf{c})$ as being made of two components $\alpha\mathbf{b}$ and $\beta\mathbf{c}$, as shown in Fig. 1-7, i.e.,

$$\mathbf{a} \times (\mathbf{b} \times \mathbf{c}) = \alpha\mathbf{b} + \beta\mathbf{c}. \tag{1.25}$$

Since

$$\mathbf{a} \cdot [\mathbf{a} \times (\mathbf{b} \times \mathbf{c})] = 0,$$

hence

$$\alpha\,(\mathbf{a} \cdot \mathbf{b}) + \beta\,(\mathbf{a} \cdot \mathbf{c}) = 0.$$

Equation (1.25), therefore, can be written in the form

$$\mathbf{a} \times (\mathbf{b} \times \mathbf{c}) = \alpha \left[\mathbf{b} - \frac{\mathbf{a} \cdot \mathbf{b}}{\mathbf{a} \cdot \mathbf{c}}\mathbf{c}\right] = \alpha' \left[(\mathbf{a} \cdot \mathbf{c})\,\mathbf{b} - (\mathbf{a} \cdot \mathbf{b})\,\mathbf{c}\right] \tag{1.26}$$

where α' is a constant to be determined. By considering the case $\mathbf{a} = \mathbf{u}_y, \mathbf{b} = \mathbf{u}_x, \mathbf{c} = \mathbf{u}_y$, we have

$$\begin{aligned} \mathbf{a} \times (\mathbf{b} \times \mathbf{c}) &= \mathbf{u}_x \\ (\mathbf{a} \cdot \mathbf{c})\,\mathbf{b} &= \mathbf{u}_x \\ (\mathbf{a} \cdot \mathbf{b})\,\mathbf{c} &= 0. \end{aligned}$$

Hence, $\alpha' = 1$. It can be shown that if the unit vectors form a left-hand system, the same result is obtained. The validity of (1.23) and (1.24) is independent of the choice of the coordinate system in which these vectors are represented.

CHAPTER 2

Coordinate Systems

2-1 ORTHOGONAL SYSTEMS

In this book, all coordinate systems used are orthogonal unless stated otherwise. The coordinate variables and the corresponding unit vectors in the general orthogonal system will be denoted by (v_1, v_2, v_3) and $(\mathbf{u}_1, \mathbf{u}_2, \mathbf{u}_3)$. The unit vectors form a right-handed system, i.e., $\mathbf{u}_1 \times \mathbf{u}_2 = \mathbf{u}_3$. A vector which corresponds to the vectorial distance measured from the origin of the coordinate system to a point P located at (v_1, v_2, v_3) will be denoted by $\mathbf{r}_p$. It is called the position vector, as shown in Fig. 2-1. If we displace P to a neighboring point Q located at $(v_1 + dv_1, v_2 + dv_2, v_3 + dv_3)$, then the total differential of $\mathbf{r}_p$ can be written in the form

$$d\mathbf{r}_p = \frac{\partial \mathbf{r}_p}{\partial v_1} dv_1 + \frac{\partial \mathbf{r}_p}{\partial v_2} dv_2 + \frac{\partial \mathbf{r}_p}{\partial v_3} dv_3. \tag{2.1}$$

The vector coefficient $\partial \mathbf{r}_p / \partial v_1$ is a measure of the change of $\mathbf{r}_p$ due to a change of v_1 only; hence, it must lie in the direction of $\mathbf{u}_1$. Thus, we can write

$$\frac{\partial \mathbf{r}_p}{\partial v_i} = h_i \, \mathbf{u}_i, \qquad i = 1, 2, 3. \tag{2.2}$$

The coefficients h_i's are called the metric coefficients of that system. If one of the coordinate variables has the dimension of length, then the corresponding metric coefficient is dimensionless. If that particular coordinate variable is dimensionless, such as an angular variable, then the corresponding metric coefficient has the dimension of length. For the Cartesian system, $(v_1, v_2, v_3) = (x, y, z)$ and $(h_1, h_2, h_3) = (1, 1, 1)$, and

$$\mathbf{r}_p = x\,\mathbf{u}_x + y\,\mathbf{u}_y + z\,\mathbf{u}_z. \tag{2.3}$$

For convenience, we will use (x_1, x_2, x_3) for (x, y, z) and $(\mathbf{a}_1, \mathbf{a}_2, \mathbf{a}_3)$ for $(\mathbf{u}_x, \mathbf{u}_y, \mathbf{u}_z)$ to simplify the writing for the summations in the Cartesian system whenever there is no confusion of notation. The metric coefficients of an orthogonal system can be found if we know the relationship between (x, y, z) and (v_1, v_2, v_3) of that system. Thus, according to (2.1) and (2.2),

$$\frac{\partial \mathbf{r}_p}{\partial v_1} = h_1\,\mathbf{u}_1. \tag{2.4}$$

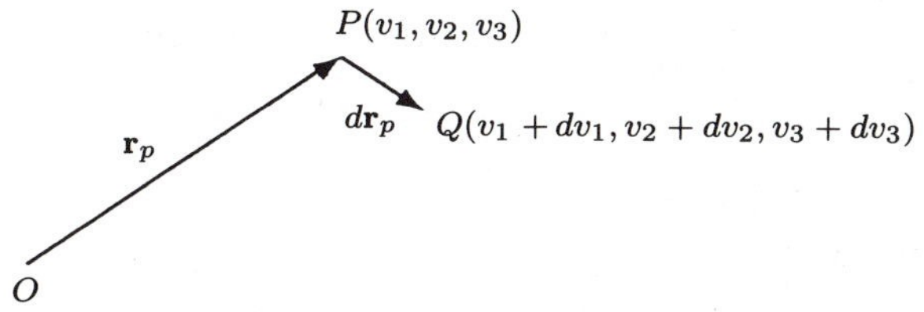

Fig. 2-1 Position vector and its total differential.

From (2.3), the partial derivation of $\mathbf{r}_p$ with respect to v_1 is

$$\frac{\partial \mathbf{r}_p}{\partial v_1} = \frac{\partial x}{\partial v_1}\,\mathbf{u}_x + \frac{\partial y}{\partial v_1}\,\mathbf{u}_y + \frac{\partial z}{\partial v_1}\,\mathbf{u}_z; \tag{2.5}$$

hence,

$$\frac{\partial \mathbf{r}_p}{\partial v_1} \cdot \frac{\partial \mathbf{r}_p}{\partial v_1} = h_1^2 = \left(\frac{\partial x}{\partial v_1}\right)^2 + \left(\frac{\partial y}{\partial v_1}\right)^2 + \left(\frac{\partial z}{\partial v_1}\right)^2.$$

In general,

$$h_i^2 = \left(\frac{\partial x}{\partial v_i}\right)^2 + \left(\frac{\partial y}{\partial v_i}\right)^2 + \left(\frac{\partial z}{\partial v_i}\right)^2, \qquad i = 1, 2, 3$$

or

$$h_i^2 = \sum_j \left(\frac{\partial x_j}{\partial v_i}\right)^2, \qquad i = 1, 2, 3. \tag{2.6}$$

The summation in (2.6) is from $j = 1$ to 3; that labeling will be omitted unless specified otherwise.

The table that follows shows the metric coefficients of some commonly used orthogonal systems, and the relationship between (v_1, v_2, v_3) and (x, y, z), based on which these metric coefficients are derived by means of (2.6).

Cartesian

coordinate variables: (x, y, z)
metric coefficients: $(1, 1, 1)$.

Cylindrical

coordinate variables: (r, ϕ, z)
metric coefficients: $(1, r, 1)$
relations: $x = r\cos\phi, y = r\sin\phi, z = z$.

Spherical

coordinate variables: (R, θ, ϕ)
metric coefficients: $(1, R, R\sin\theta)$
relations: $x = R\sin\theta\cos\phi, y = R\sin\theta\sin\phi, z = R\cos\theta$.

Elliptical Cylinder

coordinate variables: (η, ξ, z)
metric coefficients:

$$\left[c\left(\frac{\xi^2-\eta^2}{1-\eta^2}\right)^{\frac{1}{2}}, c\left(\frac{\xi^2-\eta^2}{\xi^2-1}\right)^{\frac{1}{2}}, 1\right]$$

relations: $x = c\eta\xi, y = c\left[\left(1-\eta^2\right)\left(\xi^2-1\right)\right]^{\frac{1}{2}}, z = z$.

Parabolic Cylinder

coordinate variables: (η, ξ, z)
metric coefficients:

$$\left[\left(\eta^2+\xi^2\right)^{\frac{1}{2}}, \left(\eta^2+\xi^2\right)^{\frac{1}{2}}, 1\right]$$

relations: $x = \frac{1}{2}\left(\eta^2-\xi^2\right), y = \eta\xi, z = z$.

Prolate Spheroidal

coordinate variables: (η, ξ, z)
metric coefficients:

$$\left[c\left(\frac{\xi^2-\eta^2}{1-\eta^2}\right)^{\frac{1}{2}}, c\left(\frac{\xi^2-\eta^2}{\xi^2-1}\right)^{\frac{1}{2}}, c\left(1-\eta^2\right)^{\frac{1}{2}}\left(\xi^2-1\right)^{\frac{1}{2}}\right]$$

relations: $x = c\left[\left(1-\eta^2\right)\left(\xi^2-1\right)\right]^{\frac{1}{2}}\cos\phi$,

$y = c\left[\left(1-\eta^2\right)\left(\xi^2-1\right)\right]^{\frac{1}{2}}\sin\phi$,

$z = c\eta\xi$.

Oblate Spheroidal

coordinate variables: (ξ, η, ϕ)
metric coefficients:

$$\left[c\left(\frac{\xi^2-\eta^2}{\xi^2-1}\right)^{\frac{1}{2}}, c\left(\frac{\xi^2-\eta^2}{1-\eta^2}\right)^{\frac{1}{2}}, c\xi\eta\right]$$

relations: $x = c\xi\eta\cos\phi$,

$$y = c\xi\eta\sin\phi, z = c\left[\left(\xi^2-1\right)\left(1-\eta^2\right)\right]^{\frac{1}{2}}.$$

Bipolar Cylinders

coordinate variables: (η, ξ, z)
metric coefficients:

$$\left[\frac{a}{\cosh\xi-\cos\eta}, \frac{a}{\cosh\xi-\cos\eta}, 1\right]$$

relations: $x = \dfrac{a\sinh\xi}{\cosh\xi-\cos\eta}$,

$$y = \frac{a\sin\eta}{\cosh\xi-\cos\eta}, z = z.$$

In the above, for the case of the elliptical cylinder, the governing equations for the elliptical cylinder and the conformal hyperbolic cylinder are

$$\frac{x^2}{c^2\xi^2} + \frac{y^2}{c^2\left(\xi^2-1\right)} = 1, \qquad \infty > \xi \geq 1 \tag{2.7}$$

$$\frac{x^2}{c^2\eta^2} - \frac{y^2}{c^2\left(1-\eta^2\right)} = 1, \qquad 1 \geq \eta \geq -1 \tag{2.8}$$

where c denotes half of the focal distance between the foci of the ellipse. For the prolate spheroid, the governing equations are

$$\frac{z^2}{c^2\xi^2} + \frac{r^2}{c^2\left(\xi^2-1\right)} = 1, \qquad 2\pi \geq \phi \geq 0, \infty > \xi \geq 1 \tag{2.9}$$

$$\frac{z^2}{c^2\eta^2} - \frac{r^2}{c^2\left(1-\eta^2\right)} = 1, \qquad 2\pi \geq \phi \geq 0, 1 \geq \eta \geq -1 \tag{2.10}$$

where $r^2 = x^2 + y^2$. Equation (2.9) represents an ellipse of revolution revolving around the z axis, which is the major axis. The conformal hyperboloid is represented by (2.10). For the oblate spheroid, the governing equations are

$$\frac{r^2}{c^2\xi^2} + \frac{z^2}{c^2\left(\xi^2-1\right)} = 1, \qquad 2\pi \geq \phi \geq 0, \infty > \xi \geq 1 \tag{2.11}$$

$$\frac{r^2}{c^2\eta^2} - \frac{z^2}{c^2\left(1-\eta^2\right)} = 1, \qquad 2\pi \geq \phi \geq 0, 1 \geq \eta \geq -1. \tag{2.12}$$

Equation (2.11) represents an oblate spheroid generated by revolving an ellipse around the z axis, which is the minor axis in this case, and (2.12) is the equation for the conformal hyperboloid.

For the bipolar cylinders, the governing equations are

$$(x - a \coth \xi)^2 + y^2 = a^2 \operatorname{csch}^2 \xi \tag{2.13}$$

$$(y - a \cot \eta)^2 + x^2 = a^2 \csc^2 \eta, \tag{2.14}$$

$\infty > \xi > -\infty$, $2\pi > \eta > 0$. a denotes half of the distance between two pivoting points from which these circles are generated. In fact, (2.13) and (2.14) can be derived by considering a conformal transformation between the complex variables $x + jy$ and $\eta + j\xi$ in the form

$$\frac{(x+jy)-a}{(x+jy)+a} = e^{j(\eta+j\xi)} \tag{2.15}$$

which is called a bilinear transformation in the theory of complex variables.

In the complex (x, y) plane, the numbers $c_1 = (x-a) + jy$ and $c_2 = (x+a) + jy$ are shown graphically in Fig. 2-2, where we assume "a" to be real. These numbers can also be written in the form

$$c_1 = \left[(x-a)^2 + y^2\right]^{\frac{1}{2}} e^{j\alpha_1}, \qquad \alpha_1 = \tan^{-1}\frac{y}{x-a}$$

$$c_2 = \left[(x+a)^2 + y^2\right]^{\frac{1}{2}} e^{j\alpha_2}, \qquad \alpha_2 = \tan^{-1}\frac{y}{x+a} \tag{2.16}$$

Equation (2.15) is, therefore, equivalent to

$$\frac{c_1}{c_2} = \left[\frac{(x-a)^2+y^2}{(x+a)^2+y^2}\right]^{\frac{1}{2}} e^{j(\alpha_1-\alpha_2)}$$

$$= e^{-\xi} e^{j\eta}; \tag{2.17}$$

hence,

$$\left[\frac{(x-a)^2+y^2}{(x+a)^2+y^2}\right]^{\frac{1}{2}} = e^{-\xi} \tag{2.18}$$

$$\tan^{-1}\frac{y}{x-a} - \tan^{-1}\frac{y}{x+a} = \eta. \tag{2.19}$$

It is not difficult to deduce (2.13) from (2.18), and (2.14) from (2.19). From this discussion, we see that the locus of constant ξ, which is a circle, corresponds to a constant ratio of the magnitude of c_1 and c_2, and the locus of constant η is also a circle conformal to the circle of constant ξ. The fact that they are conformal is because (2.15) is a conformal transformation in the theory of complex numbers.

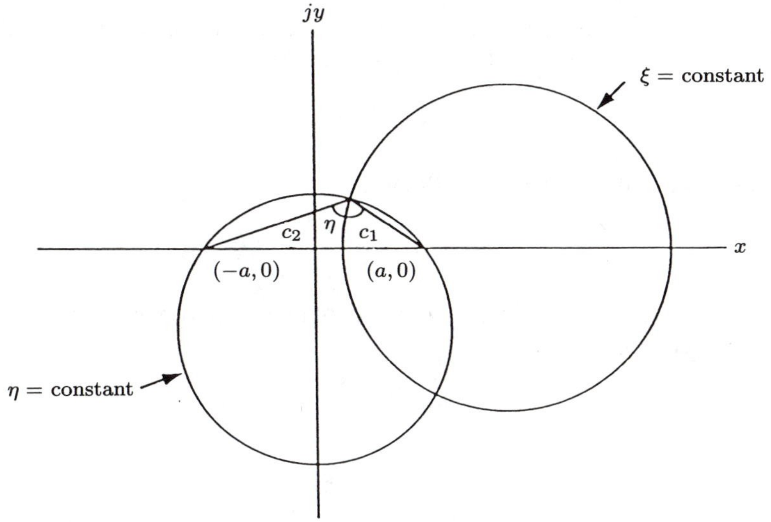

Fig. 2-2 Locus of the complex number $x + jy$ resulting from a bilinear transformation.

2-2 DERIVATIVES OF UNIT VECTORS

Equation (2.2) states

$$\frac{\partial \mathbf{r}_p}{\partial v_i} = h_i \mathbf{u}_i, \qquad i = 1, 2, 3. \tag{2.20}$$

Let us consider two different variables v_j and v_k; then

$$\frac{\partial \mathbf{r}_p}{\partial v_j} = h_j\, \mathbf{u}_j, \qquad \frac{\partial \mathbf{r}_p}{\partial v_k} = h_k\, \mathbf{u}_k.$$

Hence,

$$\frac{\partial\,(h_j\, \mathbf{u}_j)}{\partial v_k} = \frac{\partial\,(h_k\, \mathbf{u}_k)}{\partial v_j} \tag{2.21}$$

because both are equal to $\partial^2 \mathbf{r}_p / \partial v_j\, \partial v_k$. We assume that all of the first and second derivatives of $\mathbf{r}_p$ do exist. Equation (2.21) can be written in the form

$$h_j \frac{\partial \mathbf{u}_j}{\partial v_k} + \frac{\partial h_j}{\partial v_k} \mathbf{u}_j = h_k \frac{\partial \mathbf{u}_k}{\partial v_j} + \frac{\partial h_k}{\partial v_j}\, \mathbf{u}_k. \tag{2.22}$$

Since

$$\mathbf{u}_j \;\cdot\; \mathbf{u}_j = 1,$$

hence

$$\mathbf{u}_j \;\cdot\; \frac{\partial \mathbf{u}_j}{\partial v_k} = 0.$$

$\partial \mathbf{u}_j/\partial v_k$ is, therefore, perpendicular to $\mathbf{u}_j$. In the (v_j, v_k) plane, it is parallel to $\mathbf{u}_k$; similarly, $\partial \mathbf{u}_k/\partial v_j$ is parallel to $\mathbf{u}_j$. Thus, we can write

$$\frac{\partial \mathbf{u}_j}{\partial v_k} = \alpha\, \mathbf{u}_k, \quad \frac{\partial \mathbf{u}_k}{\partial v_j} = \beta \mathbf{u}_j.$$

Equation (2.22) can now be put in the form

$$\left(\frac{\partial h_k}{\partial v_j} - \alpha\, h_j\right) \mathbf{u}_k = \left(\frac{\partial h_j}{\partial v_k} - \beta\, h_k\right) \mathbf{u}_j.$$

Since $\mathbf{u}_k$ and $\mathbf{u}_j$ are independent and orthogonal to each other, the above equation can be satisfied only if

$$\frac{\partial h_k}{\partial v_j} = \alpha\, h_j$$

$$\frac{\partial\, h_j}{\partial v_k} = \beta\, h_k;$$

hence,

$$\frac{\partial \mathbf{u}_j}{\partial v_k} = \frac{1}{h_j} \frac{\partial h_k}{\partial v_j} \mathbf{u}_k \tag{2.23}$$

and

$$\frac{\partial \mathbf{u}_k}{\partial v_j} = \frac{1}{h_k} \frac{\partial h_j}{\partial v_k} \mathbf{u}_j. \tag{2.24}$$

Equations (2.23) and (2.24) hold true for j and $k = 1, 2, 3$ with $j \neq k$.

The derivative $\partial \mathbf{u}_i/\partial v_i$ can be found by considering the relationship among the three orthogonal unit vectors $\mathbf{u}_i$, $\mathbf{u}_j$, and $\mathbf{u}_k$ in a right-hand system:

$$\mathbf{u}_i = \mathbf{u}_j \times \mathbf{u}_k, \qquad i, j, k = \begin{cases} 1, & 2, & 3 & \text{or} \\ 2, & 3, & 1 & \text{or} \\ 3, & 1, & 2. \end{cases} \tag{2.25}$$

The coordinate variables of the system are denoted by (v_1, v_2, v_3). The partial derivative of (2.25) with respect to v_i yields

$$\frac{\partial \mathbf{u}_i}{\partial v_i} = \mathbf{u}_j \times \frac{\partial \mathbf{u}_k}{\partial v_i} + \frac{\partial \mathbf{u}_j}{\partial v_i} \times \mathbf{u}_k.$$

In view of (2.23) or (2.24), the above equation is equivalent to

$$\begin{aligned} \frac{\partial \mathbf{u}_i}{\partial v_i} &= \mathbf{u}_j \times \frac{1}{h_k} \frac{\partial h_i}{\partial v_k} \mathbf{u}_i + \frac{1}{h_j} \frac{\partial h_i}{\partial v_j} \mathbf{u}_i \times \mathbf{u}_k \\ &= -\left(\frac{1}{h_k} \frac{\partial h_i}{\partial v_k} \mathbf{u}_k + \frac{1}{h_j} \frac{\partial h_i}{\partial v_j} \mathbf{u}_j \right). \end{aligned} \tag{2.26}$$

Equations (2.23) and (2.26) are very important formulas which will be used frequently in subsequent sections. It can be easily verified that as a result of (2.26),

$$\sum_{i=1}^{3} \frac{\partial}{\partial v_i} \left(\frac{\Omega}{h_i} \mathbf{u}_i \right) = 0 \tag{2.27}$$

where $\Omega = h_1 \, h_2 \, h_3$ or, more explicitly,

$$\frac{\partial}{\partial v_1} (h_2 \, h_3 \, \mathbf{u}_1) + \frac{\partial}{\partial v_2} (h_1 \, h_3 \, \mathbf{u}_2) + \frac{\partial}{\partial v_3} (h_1 \, h_2 \, \mathbf{u}_3) = 0.$$

Another identity which can be proved with the aid of (2.23) and (2.26) is

$$\sum_i \frac{\mathbf{u}_i}{h_i} \times \frac{\partial}{\partial v_i} \left(\frac{\mathbf{u}_j}{h_j} \right) = 0, \qquad j = 1, 2, 3. \tag{2.28}$$

Equations (2.27) and (2.28) are needed in the derivation of many important formulas. The interpretation of these two equations from the point of view of vector theorems and identities will be discussed in Chapter 4.

It should be mentioned that the relations described by (2.23) and (2.26) have previously been derived by Morse and Feshbach [10, pp. 25–26] using both an algebraic and a geometrical approach. In comparison, the derivation provided here appears to be much simpler.

2-3 DUPIN COORDINATE SYSTEM

The Dupin coordinate system is an indispensable tool to treat vector analysis on a surface. In the general Dupin system, the coordinate variables will be denoted by (v_1, v_2, v_3) and the corresponding unit vectors by $(\mathbf{u}_1, \mathbf{u}_2, \mathbf{u}_3)$ with metric coefficients $(h_1, h_2, 1)$. The variables (v_1, v_2) are used to describe the coordinate lines on the surface, while v_3 denotes the normal distance measured linearly from the surface; hence, $h_3 = 1$. For a right-hand system, the direction of $\mathbf{u}_3$ is determined by $\mathbf{u}_1 \times \mathbf{u}_2$. $\mathbf{u}_1$ and $\mathbf{u}_2$ are orthogonal, and both are tangential to the surface. Figure 2-3 shows the disposition of these quantities. The total differential of the position vector measured from a point on the surface to a neighboring point in the space is then written as

$$d\,\mathbf{r}_p = h_1 \, dv_1 \, \mathbf{u}_1 + h_2 \, dv_2 \, \mathbf{u}_2 + dv_3 \, \mathbf{u}_3. \tag{2.29}$$

When v_1 and v_2 have not yet been specified, we designate the system as the general Dupin system. The surface of a circular cylinder and that of a sphere are two of the simplest surfaces belonging to the Dupin system. The one-to-one correspondence of the variables, the unit vectors, and the metric coefficients is listed in Table 2-1.

Let us now consider a spheroidal surface described by

$$\frac{x^2 + y^2}{b^2} + \frac{z^2}{a^2} = 1$$

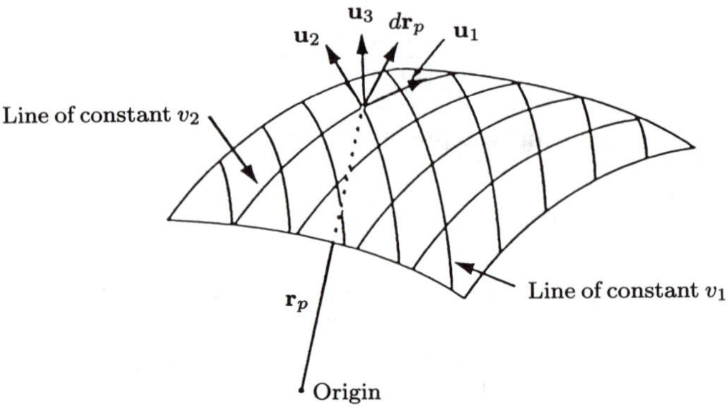

Fig. 2-3 Dupin coordinate system, $d\mathbf{r}_p = h_1 dv_1 \mathbf{u}_1 + h_2 dv_2 \mathbf{u}_2 + dv_3 \mathbf{u}_3$.

TABLE 2-1

System	(v_1, v_2, v_3)	$(\mathbf{u}_1, \mathbf{u}_2, \mathbf{u}_3)$	(h_1, h_2, h_3)
Cylindrical surface	(ϕ, z, r)	$(\mathbf{u}_\phi, \mathbf{u}_z, \mathbf{u}_r)$	$(r, 1, 1)$
Spherical surface	(θ, ϕ, R)	$(\mathbf{u}_\theta, \mathbf{u}_\phi, \mathbf{u}_R)$	$(R, R\sin\theta, 1)$

or

$$\frac{r^2}{b^2} + \frac{z^2}{a^2} = 1, \qquad 2\pi \geq \phi \geq 0 \tag{2.30}$$

where (ϕ, z, r) are the cylindrical variables. If we choose (z, ϕ) as (v_1, v_2), then

$$d\mathbf{r}_p = ds_1\, \mathbf{u}_1 + ds_2\, \mathbf{u}_2 + ds_3\, \mathbf{u}_3 \tag{2.31}$$

where

$$\begin{aligned} ds_1 &= \left[(dr)^2 + (dz)^2\right]^{\frac{1}{2}} \\ &= \left[1 + \left(\frac{dr}{dz}\right)^2\right]^{\frac{1}{2}} dz = h_1\, dv_1 \\ ds_2 &= r\, d\phi = b\left[1 - \left(\frac{z}{a}\right)^2\right]^{\frac{1}{2}} d\phi = h_2\, dv_2 \\ ds_3 &= dv_3. \end{aligned}$$

Hence,

$$\begin{aligned} h_1 &= \left[1 + \left(\frac{dr}{dz}\right)^2\right]^{\frac{1}{2}} = \sec\, \alpha \\ h_2 &= b\left[1 - \left(\frac{z}{a}\right)^2\right]^{\frac{1}{2}} \end{aligned}$$

$$\frac{dr}{dz} = \tan\,\alpha = \text{ slope of the tangent to the ellipse, (2.30), making an angle } \alpha \text{ with the } z \text{ axis.}$$

The corresponding unit vectors are

$$\mathbf{u}_1 = \sin\alpha\,\mathbf{u}_r + \cos\alpha\,\mathbf{u}_z \tag{2.32}$$

$$\mathbf{u}_2 = \mathbf{u}_\phi \tag{2.33}$$

$$\mathbf{u}_3 = \cos\alpha\,\mathbf{u}_r - \sin\alpha\,\mathbf{u}_z. \tag{2.34}$$

The choice of (v_1, v_2) in a Dupin system is not unique. In the above example, we can use (r, ϕ) as (v_1, v_2); then

$$ds_1 = \sqrt{1 + \left(\frac{dz}{dr}\right)^2}\, dr = \csc\alpha\, dr = h_1\, dv_1. \tag{2.35}$$

Hence, $h_1 = \csc\alpha$, but h_2, $\mathbf{u}_1$, and $\mathbf{u}_2$ remain the same. In the case of a spherical surface, we can use (x, z) as (v_1, v_2). In certain problems dealing with integrations, such a choice sometimes is desirable, particularly from the point of view of numerical calculations.

2-4 RADII OF CURVATURE

For a surface described in the general Dupin coordinate system, there are two radii of curvature of the surface associated with two contours in the (v_1, v_3) plane and the (v_2, v_3) plane. These are two normal planes containing $(\mathbf{u}_1, \mathbf{u}_3)$ and $(\mathbf{u}_2, \mathbf{u}_3)$ at $P\,(v_1, v_2, v_3)$ (see Fig. 2-3). These radii of curvature are closely related to the metric coefficients h_1, h_2 and the rate of change of these coefficients with respect to v_3. Figure 2-4 shows a section of the contour c_1 in the neighborhood of P resulting from the intersection of the $(\mathbf{u}_1, \mathbf{u}_3)$ plane with the surface.

Referring to Fig. 2-4, we denote

$$PQ = h_1\, dv_1$$

$$OP = R_1 \text{ (first principal radius of the curvature)}$$

$$PS = QT = dv_3.$$

Then

$$ST = \left(h_1 + \frac{\partial h_1}{\partial v_3} dv_3\right)\, dv_1$$

$$Q'Q = PQ - ST = -\frac{\partial h_1}{\partial v_3} dv_3\, dv_1.$$

The triangles OPQ and $TQ'Q$ are similar; hence,

$$\frac{PQ}{OP} = \frac{Q'Q}{QT}, \tag{2.36}$$

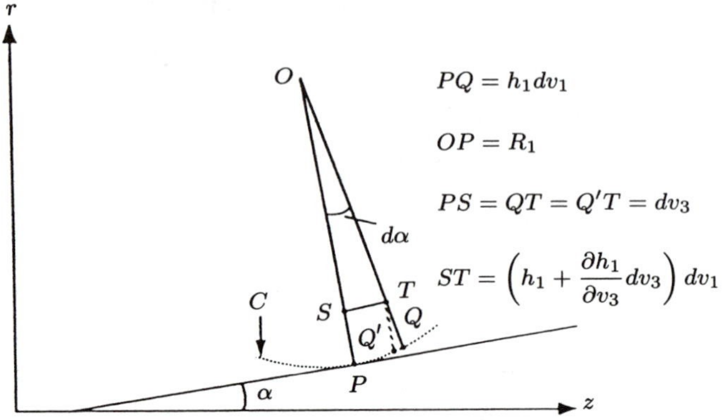

Fig. 2-4 Radius of curvature of a surface in the plane containing $\mathbf{u}_1$ and $\mathbf{u}_3$ ($r - z$ plane for the example illustrated in the text).

which yields

$$\frac{h_1}{R_1} = -\frac{\partial h_1}{\partial v_3} \tag{2.37}$$

or

$$\frac{1}{R_1} = -\frac{1}{h_1}\frac{\partial h_1}{\partial v_3}. \tag{2.38}$$

Equation (2.38) relates the first principal radius of curvature in the (v_1, v_3) plane at P to the metric coefficient h_1 and its derivative with respect to v_3 at that point. To find the expression of R_1 in terms of the shape of the curve, we have to know the governing equation for c_1. Let this equation be given in the form

$$r = f_1(z) \tag{2.39}$$

where z represents v_1; then

$$\begin{aligned} ds_1 &= \sqrt{(dr)^2 + (dz)^2} = \sqrt{1 + \left(\frac{dr}{dz}\right)^2}\, d\,z \\ &= h_1\, d\, v_1 \\ h_1 &= \sqrt{1 + \left(\frac{dr}{dz}\right)^2} = \sqrt{1 + \tan^2\, \alpha} = \sec\, \alpha \end{aligned}$$

where α denotes the angle of inclination of the tangent at P made with the z axis. Now,

$$d\alpha = \frac{h_1 dv_1}{R_1} = \frac{h_1 dz}{R_1};$$

hence,

$$\frac{1}{R_1} = \frac{1}{h_1}\frac{d\alpha}{dz} = \cos\alpha \frac{d\alpha}{dz}.$$

Since

$$\tan\,\alpha = \frac{dr}{dz} = r',$$

a differentiation of the above equation with respect to z yields

$$\frac{1}{\cos^2\alpha}\,\frac{d\alpha}{dz} = \frac{dr'}{dz} = r''.$$

We have

$$\frac{1}{R_1} = \left(\cos^3\,\alpha\right)\,r'' = \frac{r''}{(1+r'^2)^{\frac{3}{2}}}. \tag{2.40}$$

The derivation of this formula is found in many books on calculus. It is repeated here to show its relationship with the derivative of the relevant metric coefficient. Equation (2.40) shows that for a concave surface, $r'' > 0$, so R_1 is positive, and for a convex surface, R_1 is negative. Similarly, one finds that in the $(\mathbf{u}_2, \mathbf{u}_3)$ plane,

$$\frac{1}{R_2} = \frac{-1}{h_2}\,\frac{\partial h_2}{\partial v_3} \tag{2.41}$$

where R_2 denotes the second principal radius of curvature. A formula similar to (2.40) can be derived if the governing equation of the curve in the (v_2, v_3) plane is known. The reciprocals of the two radii of curvature are called Gaussian curvatures of the surface in the two orthogonal planes.

As an illustration of the application of these formulas, we consider the equation of a paraboloidal surface defined by

$$r^2 = 4fz, \qquad 2\pi \geq \phi \geq 0 \tag{2.42}$$

where f denotes the focal length of the paraboloid and (r, ϕ, z) denote the cylindrical variables. The coordinates in the Dupin system for the surface are identified as $(v_1, v_2, v_3) = (z, \phi, v_3)$ with

$$h_1 = \sqrt{1+\left(\frac{dr}{dz}\right)^2} = \sqrt{1+r'^2}, \qquad h_2 = r,\ h_3 = 1.$$

For the surface under consideration

$$r' = \frac{dr}{dz} = 2\sqrt{\frac{f}{z}},\ r'' = -\frac{1}{2}\sqrt{f}\,z^{-\frac{3}{2}}.$$

Upon substituting r' and r'' into (2.40), we find

$$R_1 = -2f\left(1+\frac{z}{f}\right)^{\frac{3}{2}}. \tag{2.43}$$

At $z = 0, R_1 = -2f$, and at $z = f, R_1 = -4\sqrt{2}f$, etc. To find R_2, it is simpler to use (2.41) instead of finding the equation of the cross section in the $(\mathbf{u}_2, \mathbf{u}_3)$ plane. Now,

$$\frac{1}{R_2} = \frac{-1}{h_2}\frac{\partial h_2}{\partial v_3} = \frac{-1}{r}\frac{\partial r}{\partial v_3} = -\frac{\sin\beta}{r} \tag{2.44}$$

where β denotes the angle between the normal to the surface $\mathbf{u}_3$ and the z axis, and

$$\tan\beta = -\frac{1}{r'},\ \sin\beta = \frac{1}{(1+r'^2)^{\frac{1}{2}}}.$$

Thus,

$$R_2 = -r\left(1+r'^2\right)^{\frac{1}{2}} = -2f\left(1+\frac{z}{f}\right)^{\frac{1}{2}}. \tag{2.45}$$

At $z = 0, R_2 = -2f$, and at $z = f, R_2 = -2\sqrt{2}f = \frac{1}{2}R_1$. The relationship between R_1 and R_2, in general, is $R_2^3 = 4f^2 R_1$.

As an exercise, the reader may be interested to verify that for a spheroidal surface defined by

$$\left(\frac{r}{b}\right)^2 + \left(\frac{z}{a}\right)^2 = 1, \qquad 2\pi \geq \phi \geq 0,$$

$$R_1 = -\frac{a^2}{b}\left[1+\frac{z^2}{a^2}\left(\frac{b^2}{a^2}-1\right)\right]^{\frac{3}{2}} \tag{2.46}$$

and

$$R_2 = -b\left[1+\frac{z^2}{a^2}\left(\frac{b^2}{a^2}-1\right)\right]^{\frac{1}{2}}. \tag{2.47}$$

The relationship between R_1 and R_2 is $R_2^3 = \left(b^4/a^2\right)R_1$.

CHAPTER 3

Line Integrals, Surface Integrals, and Volume Integrals

3-1 DIFFERENTIAL LENGTH, AREA, AND VOLUME

In this section, we shall give a brief review of the differential quantities to be used in vector analysis, particularly, their notation. A differential length, in general, will be denoted by $d\boldsymbol{\ell}$. It is the same as the *total differential* of the position vector $d\mathbf{r}_p$, which is given by (2.1)–(2.2); hence,

$$d\boldsymbol{\ell} = \sum_i h_i \, dv_i \, \mathbf{u}_i. \tag{3.1}$$

For a cell with its center located at (v_1, v_2, v_3) and bounded by six surfaces located at $v_i \pm dv_i/2$ with $i = 1, 2, 3$, the *vector differential area* of the three surfaces located at $v_i \pm dv_i/2$ is then given by

$$\begin{aligned} d\mathbf{S}_i &= h_j \, dv_j \, \mathbf{u}_j \times h_k \, dv_k \, \mathbf{u}_k \Big|_{v_i + dv_i/2} \\ &= h_j \, h_k \, dv_j \, dv_k \, \mathbf{u}_i \Big|_{v_i + dv_i/2} \end{aligned} \tag{3.2}$$

where (i, j, k) follows the cyclic order of $(1, 2, 3)$, $(2, 3, 1)$, and $(3, 1, 2)$. The vector differential areas of the other three surfaces are

$$d\mathbf{S}_i = -h_j \, h_k \, dv_j \, dv_k \, \mathbf{u}_i \Big|_{v_i - dv_i/2} \, . \tag{3.3}$$

All of these vector areas are pointed away or outward from their surfaces. We should emphasize the fact that the metric coefficients and the corresponding unit vectors are evaluated at the sites of these surfaces $v_i \pm dv_i/2$, not at the center of the cell. The differential volume of the cell is given by

$$\begin{aligned} dv &= h_i\, dv_i\, \mathbf{u}_i \cdot (h_j\, dv_j\, \mathbf{u}_j \times h_k\, dv_k\, \mathbf{u}_k) \\ &= h_i\, h_j\, h_k\, dv_i\, dv_j\, dv_k \\ &= h_1\, h_2\, h_3\, dv_1\, dv_2\, dv_3. \end{aligned} \tag{3.4}$$

3-2 CLASSIFICATION OF LINE INTEGRALS

If we continuously change the position vector of a point in space in a certain specified manner, the locus of the point will trace a curve in space (Fig. 3-1). Let a typical point on the curve be denoted by $P\,(x, y, z)$ in the Cartesian coordinate system. If (x, y, z) are functions of a single parameter t, then as t varies, $x\,(t)\,, y\,(t)$, and $z\,(t)$ will vary accordingly. We call such a description the parametric representation of a curve. We assume that there is a one-to-one correspondence between t and (x, y, z). The vector differential length of the curve can now be written as

$$\begin{aligned} d\boldsymbol{\ell} &= dx\,\mathbf{u}_x + dy\,\mathbf{u}_y + dz\,\mathbf{u}_z \\ &= \left(\frac{dx}{dt}\mathbf{u}_x + \frac{dy}{dt}\mathbf{u}_y + \frac{dz}{dt}\mathbf{u}_z\right) dt. \end{aligned} \tag{3.5}$$

It should be pointed out here that we use dx, dy, dz, and dt to denote the total differential of these variables, but dx/dt, dy/dt, and dz/dt are the derivatives of x, y, z with respect to t. As an example, let

$$\left.\begin{aligned} x &= a\ \cos\frac{2\pi}{T}t \\ y &= a\ \sin\frac{2\pi}{T}t \\ z &= \frac{b}{T}t \end{aligned}\right\} \tag{3.6}$$

where a, b, T are constants and t is the parameter. As t varies, the locus of P describes a right-hand spiral advancing in the positive z direction as t is increased. When t changes from 0 to T, the spiral starts at $(a, 0, 0)$ and ends at $(a, 0, b)$; therefore, "b" denotes the height of the spiral after making one complete turn, and "a" denotes the radius of the circular projection of the spiral on the $x - y$ plane. To calculate the length of the spiral, one starts with

$$(d\ell)^2 = d\boldsymbol{\ell} \cdot d\boldsymbol{\ell} = \left[\left(\frac{dx}{dt}\right)^2 + \left(\frac{dy}{dt}\right)^2 + \left(\frac{dz}{dt}\right)^2\right](dt)^2;$$

hence,

$$d\ell = \left[\left(\frac{dx}{dt}\right)^2 + \left(\frac{dy}{dt}\right)^2 + \left(\frac{dz}{dt}\right)^2\right]^{\frac{1}{2}} dt. \tag{3.7}$$

The integral of (3.7) from $t = 0$ to T yields

$$L = \int_0^L d\ell = \frac{1}{T}\int_0^T \left[(2\pi a)^2 + b^2\right]^{\frac{1}{2}} dt = \left[(2\pi a)^2 + b^2\right]^{\frac{1}{2}}$$

$$= \left(c^2 + b^2\right)^{\frac{1}{2}} \tag{3.8}$$

where c denotes the circumference of the projected circle. The pitch angle of the spiral is defined by

$$\alpha = \tan^{-1}\left(\frac{b}{c}\right). \tag{3.9}$$

Equation (3.8) represents the simplest form of a line integral. In general, we define the line integral of Type I as

$$I_1 = \int_c f(x, y, z)\, d\ell \tag{3.10}$$

where c denotes the contour of the curve wherein the integration is executed.

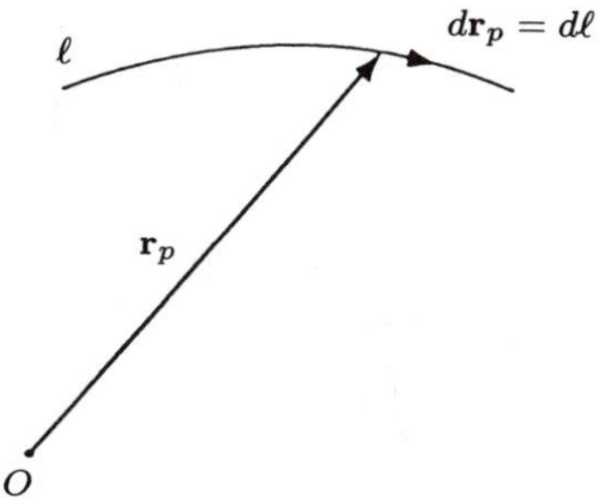

Fig. 3-1 Curve in a three-dimensional space.

As an example, let c be a parabolic curve described by

$$y^2 = 2x \text{ in the plane } z = 0, \tag{3.11}$$

and the contour extends from $x = 0, y = 0$ to $x = \frac{3}{2}, y = \sqrt{3}$, and the function in (3.10) is supposed to be $f(x, y) = xy$.

If we choose y as the parameter, then

$$\begin{aligned} d\boldsymbol{\ell} &= dx\,\mathbf{u}_x + dy\,\mathbf{u}_y \\ &= y\,dy\,\mathbf{u}_x + dy\,\mathbf{u}_y \\ d\ell &= \left(y^2 + 1\right)^{\frac{1}{2}} dy \\ f(x, y) &= \frac{1}{2}y^3; \end{aligned}$$

thus,

$$I_1 = \int_c f(x,y)\,d\ell = \int_0^{\sqrt{3}} \frac{1}{2} y^3 \left(y^2+1\right)^{\frac{1}{2}} dy$$
$$= \frac{28}{15}.$$

We purposely choose $f(x,y)$ and c in such a manner that the integral can be evaluated in a closed form in order to clearly illustrate the steps.

The second type of line integral is defined by

$$\mathbf{I}_2 = \int_c \mathbf{f}(x,y,z)\,d\ell \tag{3.12}$$

where **f** is a vector function.

If we write **f** in its component form in the Cartesian coordinate system,

$$\mathbf{f}(x,y,z) = f_x(x,y,z)\,\mathbf{u}_x + f_y(x,y,z)\,\mathbf{u}_y + f_z(x,y,z)\,\mathbf{u}_z, \tag{3.13}$$

and since $\mathbf{u}_x, \mathbf{u}_y, \mathbf{u}_z$ are constant vectors, we can change (3.12) into the form

$$\mathbf{I}_2 = \mathbf{u}_x \int_c f_x d\ell + \mathbf{u}_y \int_c f_y d\ell + \mathbf{u}_z \int f_z d\ell. \tag{3.14}$$

The three integrals contained in (3.14) are of Type I, which can be evaluated according to the method described previously. Line integrals of Types III, IV, and V are defined by

$$\mathbf{I}_3 = \int_c f(x,y,z)\,d\boldsymbol{\ell} \tag{3.15}$$

$$I_4 = \int_c \mathbf{f}(x,y,z)\cdot d\boldsymbol{\ell} \tag{3.16}$$

$$\mathbf{I}_5 = \int_c \mathbf{f}(x,y,z)\times d\boldsymbol{\ell}. \tag{3.17}$$

Integrals of Type III can be resolved into three integrals, i.e.,

$$\mathbf{I}_3 = \mathbf{u}_x \int_c f\;dx + \mathbf{u}_y \int f\;dy + \mathbf{u}_z \int_c f\;dz. \tag{3.18}$$

The three scalar integrals in (3.18) can be evaluated by choosing a proper parameter for each integral. In fact, one can, for example, use x as a parameter for the first integral and express both y and z in terms of x. In the case of the spiral contour, if we let z be the parameter, then

$$x = a\cos\frac{2\pi}{b}z$$
$$y = a\sin\frac{2\pi}{b}z.$$

Integrals of Type IV can be converted to

$$I_4 = \int_c (f_x\,dx + f_y\,dy + f_z\,dz)\,. \tag{3.19}$$

Again, each term in (3.19) can be evaluated by the parametric method. Integrals of Type V can be written in the form

$$\mathbf{I}_5 = \mathbf{u}_x \int_c (f_2\,dz - f_3\,dy) + \mathbf{u}_y \int (f_3\,dx - f_1 dy) + \mathbf{u}_z \int_c (f_1\,dy - f_2\,dx)\,. \tag{3.20}$$

All the six terms in (3.20) can be evaluated in the same way. Thus, if the functions f and $\mathbf{f}$ and the differential lengths $d\ell$ and $\mathbf{d}\boldsymbol{\ell}$ are expressed in a Cartesian system, we have a systematic method to evaluate all different types of line integrals. In many cases, the curve under consideration may correspond to the intersection of two surfaces represented by

$$z = F_1(x, y) \tag{3.21}$$

$$z = F_2(x, y)\,. \tag{3.22}$$

In that case, we can eliminate z between (14) and (15) so that

$$F_1(x, y) = F_2(x, y) \tag{3.23}$$

and then solve for x in terms of y to yield

$$x = F_3(y) \tag{3.24}$$

$$z = F_1[F_3(y), y]\,. \tag{3.25}$$

It is obvious that y can be used as the parameter for the curve. In many problems, it is sometimes rather difficult to find the explicit form of F_3 unless F_1 and F_2 are relatively simple functions.

If the integrals and the contour c are described in an orthogonal system other than a Cartesian system, then

$$d\ell = \left[\sum_{i=1}^{3} (h_i\,dv_i)^2\right]^{\frac{1}{2}} \tag{3.26}$$

and

$$\mathbf{d}\boldsymbol{\ell} = \sum_{i=1}^{3} h_i\,dv_i\,\mathbf{u}_i\,. \tag{3.27}$$

A scalar function f is then assumed to be a function of (v_1, v_2, v_3), and a vector function $\mathbf{f}$ would be a function of both the v_i's and $\mathbf{u}_i$'s. Integrals of Types I and IV can be evaluated by expressing the v_i's and h_i's in terms of a single parameter, as was done previously. The integrands of integrals of Types II, III, and V contain $\mathbf{u}_i$'s which are, in general, not constant vectors so they cannot be removed to the outside of these integrals. For these cases, we can transform the $\mathbf{u}_i$'s in terms of $\mathbf{u}_x$, $\mathbf{u}_y$, and $\mathbf{u}_z$ in the form

$$\mathbf{u}_i = \cos\ \alpha_i\ \mathbf{u}_x + \cos\ \beta_i\ \mathbf{u}_y + \cos\ \gamma_i\ \mathbf{u}_z, \qquad i = 1, 2, 3. \tag{3.28}$$

Then

$$\begin{aligned} \mathbf{f} &= \sum_{i=1}^{3} f_i\ \mathbf{u}_i \\ &= \sum_{i=1}^{3} f_i\left(\cos\ \alpha_i\ \mathbf{u}_x + \cos\ \beta_i\ \mathbf{u}_y + \cos\ \gamma_i\ \mathbf{u}_z\right) \\ &= f_x\ \mathbf{u}_x + f_y\ \mathbf{u}_y + f_z\ \mathbf{u}_z \end{aligned} \tag{3.29}$$

where

$$f_x = \sum_{i=1}^{3} f_i\ \cos\ \alpha_i \tag{3.30}$$

$$f_y = \sum_{i=1}^{3} f_i\ \cos\ \beta_i \tag{3.31}$$

$$f_z = \sum_{i=1}^{3} f_i\ \cos\ \gamma_i. \tag{3.32}$$

Afterwards, the unit vectors $\mathbf{u}_x$, $\mathbf{u}_y$, and $\mathbf{u}_z$ can be removed to the outside of the integrals, and the remaining scalar integrals can be evaluated by the parametric method. In a later section, we will introduce a relatively simple method to find the transformation of the unit vectors from one system to another like (3.28).

3-3 CLASSIFICATION OF SURFACE INTEGRALS

A surface in a three-dimensional space, in general, is characterized by a governing equation

$$F\left(x, y, z\right) = 0 \tag{3.33}$$

in which we can select any two variables as independent and the remaining one will be a dependent variable. We assume that we can convert (3.33) into the explicit form

$$S: \qquad z = f\left(x, y\right). \tag{3.34}$$

For two neighboring points located on S, the total differential of the displacement vector between the two adjacent points can be written in the form

$$d\mathbf{r}_p = dx\ \mathbf{u}_x + dy\ \mathbf{u}_y + dz\ d\mathbf{u}_z. \tag{3.35}$$

Only two of the Cartesian variables are independent because of the constraint stated by (3.34). If the same surface can be described by the coordinates (v_1, v_2, v_3) with unit vectors $(\mathbf{u}_1, \mathbf{u}_2, \mathbf{u}_3)$ and metric coefficients $(h_1, h_2, 1)$ in a Dupin system, then the total differential of a displacement on the surface can be written as (Fig. 3-2)

$$d\mathbf{r}_p = h_1\, dv_1\, \mathbf{u}_1 + h_2\, dv_2\, \mathbf{u}_2 \tag{3.36}$$

with $dv_3 = 0$.

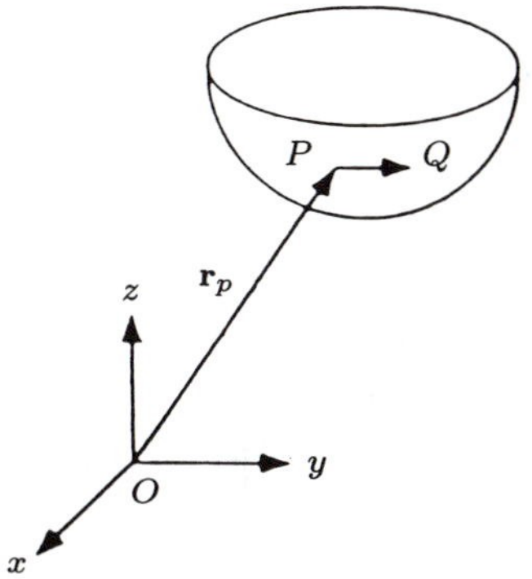

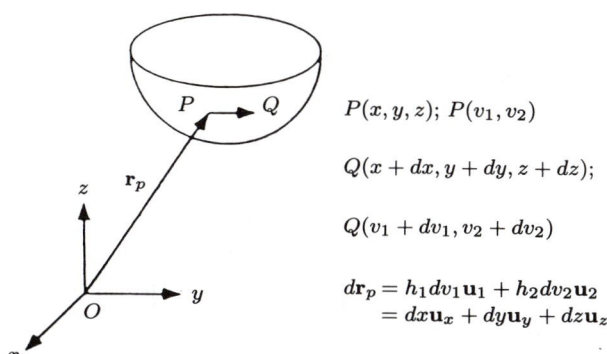

Fig. 3-2 Total differential of the position vector on a surface $z = f(x, y)$ where $v_3 =$ constant.

The partial derivatives of (3.35) and (3.36) with respect to v_1 and v_2 are

$$\frac{\partial \mathbf{r}_p}{\partial v_1} = h_1\mathbf{u}_1 = \frac{\partial x}{\partial v_1}\mathbf{u}_x + \frac{\partial y}{\partial v_1}\mathbf{u}_y + \frac{\partial z}{\partial v_1}\mathbf{u}_z \tag{3.37}$$

$$\frac{\partial \mathbf{r}_p}{\partial v_2} = h_2\mathbf{u}_2 = \frac{\partial x}{\partial v_2}\mathbf{u}_x + \frac{\partial y}{\partial v_2}\mathbf{u}_y + \frac{\partial z}{\partial v_2}\mathbf{u}_z. \tag{3.38}$$

Let the vector differential area of the surface be denoted by $d\mathbf{S}$; then

$$d\mathbf{S} = h_1\, dv_1\, \mathbf{u}_1 \times h_2\, dv_2\, \mathbf{u}_2 = h_1\, h_2\, dv_1\, dv_2\, \mathbf{u}_3$$

$$= \begin{vmatrix} \mathbf{u}_x & \mathbf{u}_y & \mathbf{u}_z \\ \dfrac{\partial x}{\partial v_1} & \dfrac{\partial y}{\partial v_1} & \dfrac{\partial z}{\partial v_1} \\ \dfrac{\partial x}{\partial v_2} & \dfrac{\partial y}{\partial v_2} & \dfrac{\partial z}{\partial v_2} \end{vmatrix} dv_1\, dv_2 = \mathbf{J}\, dv_1\, dv_2. \tag{3.39}$$

The determinant in (3.39), denoted by $\mathbf{J}$, results from $h_1\, \mathbf{u}_1 \times h_2\, \mathbf{u}_2$. For convenience, we will call it the vector Jacobian of transformation between (x, y, z) and (v_1, v_2). If we write

$$\mathbf{J} = J_x\, \mathbf{u}_x + J_y\, \mathbf{u}_y + J_z\, \mathbf{u}_z, \tag{3.40}$$

then

$$J_x = \begin{vmatrix} \dfrac{\partial y}{\partial v_1} & \dfrac{\partial z}{\partial v_1} \\ \dfrac{\partial y}{\partial v_2} & \dfrac{\partial z}{\partial v_2} \end{vmatrix} \tag{3.41}$$

is the Jacobian of transformation between (y, z) and (v_1, v_2). Sometimes, it is denoted by

$$J_x = \frac{\partial (y, z)}{\partial (v_1, v_2)}. \tag{3.42}$$

Similarly,

$$J_y = \frac{\partial (z, x)}{\partial (v_1, v_2)} \tag{3.43}$$

and

$$J_z = \frac{\partial (x, y)}{\partial (v_1, v_2)}. \tag{3.44}$$

Now, let us consider the case where the Cartesian variables (x, y) are selected as (v_1, v_2); then

$$\mathbf{J}_1 = \begin{vmatrix} \mathbf{u}_x & \mathbf{u}_y & \mathbf{u}_z \\ 1 & 0 & \dfrac{\partial z}{\partial x} \\ 0 & 1 & \dfrac{\partial z}{\partial y} \end{vmatrix}$$

$$= -\frac{\partial z}{\partial x}\mathbf{u}_x - \frac{\partial z}{\partial y}\mathbf{u}_y + \mathbf{u}_z. \tag{3.45}$$

The subscript "1" attached to $\mathbf{J}_1$ means that this is our first choice or first case. From (3.45), we can determine the unit normal vector $\mathbf{u}_3$, namely,

$$\mathbf{u}_3 = \frac{\mathbf{J}_1}{|\mathbf{J}_1|} = \frac{-\frac{\partial z}{\partial x}\mathbf{u}_x - \frac{\partial z}{\partial y}\mathbf{u}_y + \mathbf{u}_z}{\left[\left(\frac{\partial z}{\partial x}\right)^2 + \left(\frac{\partial z}{\partial y}\right)^2 + 1\right]^{\frac{1}{2}}}. \tag{3.46}$$

The directional cosines of $\mathbf{u}_3$ are, therefore, given by

$$\cos\,\alpha_3 = \frac{-\frac{\partial z}{\partial x}}{\left[\left(\frac{\partial z}{\partial x}\right)^2 + \left(\frac{\partial z}{\partial y}\right)^2 + 1\right]^{\frac{1}{2}}} \tag{3.47}$$

$$\cos\,\beta_3 = \frac{-\frac{\partial z}{\partial x}}{\left[\left(\frac{\partial z}{\partial x}\right)^2 + \left(\frac{\partial z}{\partial y}\right)^2 + 1\right]^{\frac{1}{2}}} \tag{3.48}$$

$$\cos\,\gamma_3 = \frac{1}{\left[\left(\frac{\partial z}{\partial x}\right)^2 + \left(\frac{\partial z}{\partial y}\right)^2 + 1\right]^{\frac{1}{2}}}. \tag{3.49}$$

Based on (3.39) and (3.45), we find

$$\begin{aligned} dS &= |d\mathbf{S}| = |\mathbf{J}_1|\, dv_1\, dv_2 \\ &= \left[\left(\frac{\partial z}{\partial x}\right)^2 + \left(\frac{\partial z}{\partial y}\right)^2 + 1\right]^{\frac{1}{2}} dx\, dy \\ &= \frac{1}{\cos\,\gamma_3} dx\, dy. \end{aligned} \tag{3.50}$$

Equation (3.50) can be used to find the area of a surface. As an example, let the surface be a portion of a parabola of revolution described by

$$S: \qquad z = \frac{1}{2}\left(x^2 + y^2\right), \qquad \frac{1}{2} \geq z \geq 0. \tag{3.51}$$

Then

$$\left[\left(\frac{\partial z}{\partial x}\right)^2 + \left(\frac{\partial z}{\partial y}\right)^2 + 1\right]^{\frac{1}{2}} = \left(x^2 + y^2 + 1\right)^{\frac{1}{2}}. \tag{3.52}$$

Hence,

$$S = \iint_{S_1} dS = \iint_{S_1} \left(x^2 + y^2 + 1\right)^{\frac{1}{2}} dx\, dy \tag{3.53}$$

where S_1 denotes the domain of integration with respect to (x, y), covering the projection of S in the $x - y$ plane. For this particular example, it is convenient to convert (3.53) into an integral with respect to the cylindrical variables r and ϕ, i.e.,

$$\begin{aligned} S &= \iint_{S_1} \left(1 + r^2\right)^{\frac{1}{2}} r\, dr\, d\phi = \int_0^{2\pi} \int_0^1 \left(1 + r^2\right)^{\frac{1}{2}} r\, dr\, d\phi \\ &= \frac{4\sqrt{2}\pi}{3}. \end{aligned} \tag{3.54}$$

Here, we have used the transformation

$$dx\, dy = \frac{\partial\,(x, y)}{\partial\,(r, \phi)} dr\, d\phi \tag{3.55}$$

where

$$\frac{\partial\,(x, y)}{\partial\,(r, \phi)} = \begin{vmatrix} \dfrac{\partial x}{\partial r} & \dfrac{\partial y}{\partial r} \\ \dfrac{\partial x}{\partial \phi} & \dfrac{\partial y}{\partial \phi} \end{vmatrix} = r \tag{3.56}$$

with

$$x = r\ \cos\ \phi, y = r\ \sin\ \phi.$$

Equation (3.55) is a special case of (3.39) when it is applied to a plane surface corresponding to the $x - y$ plane.

Returning now to the expression for **J** defined by (3.39), we can select either (y, z) or (z, x) as two alternative choices for (v_1, v_2); then

$$\mathbf{J}_2 = \begin{vmatrix} \mathbf{u}_x & \mathbf{u}_y & \mathbf{u}_z \\ \dfrac{\partial x}{\partial y} & 1 & 0 \\ \dfrac{\partial x}{\partial z} & 0 & 1 \end{vmatrix}$$

$$= \mathbf{u}_x - \frac{\partial x}{\partial y}\mathbf{u}_y - \frac{\partial x}{\partial z}\mathbf{u}_z. \tag{3.57}$$

In this case, x is the dependent variable. The expression for the unit vector $\mathbf{u}_3$ is now given by

$$\mathbf{u}_3 = \frac{\mathbf{J}_2}{|\mathbf{J}_2|} = \frac{\mathbf{u}_x - \frac{\partial x}{\partial y}\mathbf{u}_y - \frac{\partial x}{\partial z}\mathbf{u}_z}{\left[1 + \left(\frac{\partial x}{\partial y}\right)^2 + \left(\frac{\partial x}{\partial z}\right)^2\right]^{\frac{1}{2}}} \tag{3.58}$$

and

$$d\mathbf{S} = \mathbf{J}_2\ dy\ dz = |\mathbf{J}_2|\ dy\ dz\ \mathbf{u}_3$$

$$= \frac{1}{\cos\ \alpha_3} dy\ dz\ \mathbf{u}_3.$$

Similarly, we have

$$\mathbf{J}_3 = \begin{vmatrix} \mathbf{u}_x & \mathbf{u}_y & \mathbf{u}_z \\ 0 & \dfrac{\partial y}{\partial z} & 1 \\ 1 & \dfrac{\partial y}{\partial x} & 0 \end{vmatrix}$$

$$= -\frac{\partial y}{\partial x}\mathbf{u}_x + \mathbf{u}_y - \frac{\partial y}{\partial z}\mathbf{u}_z \tag{3.59}$$

$$\mathbf{u}_3 = \frac{\mathbf{J}_3}{|\mathbf{J}_3|} = \frac{-\frac{\partial y}{\partial x}\mathbf{u}_x + \mathbf{u}_y - \frac{\partial y}{\partial z}\mathbf{u}_z}{\left[\left(\frac{\partial y}{\partial x}\right)^2 + \left(\frac{\partial y}{\partial z}\right)^2 + 1\right]^{\frac{1}{2}}} \tag{3.60}$$

$$d\mathbf{S} = \mathbf{J}_3\ dx\ dz = |\mathbf{J}_3|\ dx\ dz\ \mathbf{u}_3$$

$$= \frac{1}{\cos\ \beta_3}\ dx\ dz\ \mathbf{u}_3. \tag{3.61}$$

The directional cosines of $\mathbf{u}_3$, therefore, can be expressed in several different forms. Our three different choices of (v_1, v_2) yield

$$\cos\ \alpha_3 = \frac{-\frac{\partial z}{\partial x}}{\left[\left(\frac{\partial z}{\partial x}\right)^2 + \left(\frac{\partial z}{\partial y}\right)^2 + 1\right]^{\frac{1}{2}}} = \frac{1}{\left[\left(\frac{\partial x}{\partial y}\right)^2 + \left(\frac{\partial x}{\partial z}\right)^2 + 1\right]^{\frac{1}{2}}}$$

$$= \frac{-\frac{\partial y}{\partial x}}{\left[\left(\frac{\partial y}{\partial x}\right)^2 + \left(\frac{\partial y}{\partial z}\right)^2 + 1\right]^{\frac{1}{2}}} \tag{3.62}$$

$$\cos\ \beta_3 = \frac{-\frac{\partial z}{\partial y}}{\left[\left(\frac{\partial z}{\partial x}\right)^2 + \left(\frac{\partial z}{\partial y}\right)^2 + 1\right]^{\frac{1}{2}}} = \frac{-\frac{\partial x}{\partial y}}{\left[\left(\frac{\partial x}{\partial y}\right)^2 + \left(\frac{\partial x}{\partial z}\right)^2 + 1\right]^{\frac{1}{2}}}$$

$$= \frac{1}{\left[\left(\frac{\partial y}{\partial x}\right)^2 + \left(\frac{\partial y}{\partial z}\right)^2 + 1\right]^{\frac{1}{2}}} \tag{3.63}$$

$$\cos\ \gamma_3 = \frac{1}{\left[\left(\frac{\partial z}{\partial x}\right)^2 + \left(\frac{\partial z}{\partial y}\right)^2 + 1\right]^{\frac{1}{2}}} = \frac{-\frac{\partial x}{\partial z}}{\left[\left(\frac{\partial x}{\partial y}\right)^2 + \left(\frac{\partial x}{\partial z}\right)^2 + 1\right]^{\frac{1}{2}}}$$

$$= \frac{-\frac{\partial y}{\partial z}}{\left[\left(\frac{\partial y}{\partial x}\right)^2 + \left(\frac{\partial y}{\partial z}\right)^2 + 1\right]^{\frac{1}{2}}}. \tag{3.64}$$

As a reminder to the beginners, we would like to call attention to the fact that for a function of single variables like $y = f(x)$, it is well known that

$$\frac{dy}{dx} = -\left(1 \bigg/ \frac{dx}{dy}\right), \tag{3.65}$$

but for functions of multiple variables,

$$\frac{\partial y}{\partial x} \neq -\left(1 \bigg/ \frac{\partial x}{\partial y}\right). \tag{3.66}$$

As an example, we consider the relations

$$x = r\ \cos\ \phi,\ \ y = r\ \sin\ \phi.$$

Then

$$r = \left(x^2 + y^2\right)^{\frac{1}{2}},\ \ \phi = \tan^{-1}\left(\frac{y}{x}\right)$$

so

$$\frac{\partial x}{\partial r} = \cos\ \phi$$

and

$$\frac{\partial r}{\partial x} = \frac{x}{(x^2 + y^2)^{\frac{1}{2}}} = \cos\,\phi,$$

that is,

$$\frac{\partial x}{\partial r} = \frac{\partial r}{\partial x},$$

while

$$\frac{\partial x}{\partial \phi} = -r\,\sin\,\phi, \quad \frac{\partial \phi}{\partial x} = -\frac{\sin\,\phi}{r}$$

so

$$\frac{\partial x}{\partial \phi} = r^2 \frac{\partial \phi}{\partial x}.$$

One, therefore, must be very careful to distinguish between the dependent and independent variables.

Like the line integrals, there are five types of surface integrals. From now on, functions of space variables (x, y, z) or (v_1, v_2, v_3) will be denoted by $F(\mathbf{r}_p)$ or $\mathbf{F}(\mathbf{r}_p)$ where $\mathbf{r}_p$ denotes the position vector. The five types of surface integrals are as follows.

Type I:

$$I_1 = \iint_S F(\mathbf{r}_p)\, dS \tag{3.67}$$

Type II:

$$\mathbf{I}_2 = \iint_S \mathbf{F}(\mathbf{r}_p)\, dS \tag{3.68}$$

Type III:

$$\mathbf{I}_3 = \iint_S F(\mathbf{r}_p)\, d\mathbf{S} \tag{3.69}$$

Type IV:

$$I_4 = \iint_S \mathbf{F}(\mathbf{r}_p) \cdot d\mathbf{S} \tag{3.70}$$

Type V:

$$\mathbf{I}_5 = \iint_S \mathbf{F}(\mathbf{r}_p) \times d\mathbf{S} \tag{3.71}$$

where $F(\mathbf{r}_p)$ is a scalar function of position and $\mathbf{F}(\mathbf{r}_p)$ denotes a vector function.

We assume that the surface S can be described by a governing equation of the form

$$z = f(x, y)\,. \tag{3.72}$$

The same surface can always be considered as a normal surface ($v_3 = 0$) in a proper Dupin system with parameters (v_1, v_2), $(\mathbf{u}_1, \mathbf{u}_2)$ and metric coefficients (h_1, h_2). Treating v_1 and v_2 as two independent variables, we can write

$$x = f_1(v_1, v_2) \tag{3.73}$$

$$y = f_2(v_1, v_2) \tag{3.74}$$

$$\begin{aligned} z &= f(x, y) = f[f_1(v_1, v_2), f_2(v_1, v_2)] \\ &= f_3(v_1, v_2). \end{aligned} \tag{3.75}$$

The functions $F(\mathbf{r}_p)$ and $\mathbf{F}(\mathbf{r}_p)$ contained in (3.67)–(3.71), therefore, can be changed into functions of (v_1, v_2) for the scalar function, and of (v_1, v_2) as well as $(\mathbf{u}_1, \mathbf{u}_2, \mathbf{u}_3)$ for the vector function. An integral of Type I can be transformed into

$$I_1 = \iint_S F(v_1, v_2) |\mathbf{J}| \, dv_1 \, dv_2 \tag{3.76}$$

which can be evaluated by the parametric method. Thus, if we let $(v_1, v_2) = (x, y)$, (3.76) becomes

$$I_1 = \iint_{S_3} F(x, y) \frac{1}{\cos \gamma_3} \, dx \, dy \tag{3.77}$$

where S_3 denotes the domain of integration on the $x - y$ plane covered by the projection of S on that plane. The execution to carry out the integration is very similar to the problem of finding the area of a curved surface, except that the integrand contains an additional function.

An integral of Type II is equivalent to

$$\begin{aligned} \mathbf{I}_2 &= \mathbf{u}_x \iint F_x(v_1, v_2) |\mathbf{J}| \, dv_1 \, dv_2 \\ &+ \mathbf{u}_y \iint F_y(v_1, v_2) |\mathbf{J}| \, dv_1 \, dv_2 \\ &+ \mathbf{u}_z \iint F_z(v_1, v_2) |\mathbf{J}| \, dv_1 \, dv_2. \end{aligned} \tag{3.78}$$

The three scalar integrals in (3.78) are of Type I. However, it is not necessary to use the same set of (v_1, v_2) for these integrals.

An integral of Type III, in view of (3.39) and (3.40), is equivalent to

$$\begin{aligned} \mathbf{I}_3 &= \iint F(v_1, v_2) \mathbf{J} \, dv_1 \, dv_2 \\ &= \mathbf{u}_x \iint J_x \, F(v_1, v_2) \, dv_1 \, dv_2 \\ &+ \mathbf{u}_y \iint J_y \, F(v_1, v_2) \, dv_1 \, dv_2 \\ &+ \mathbf{u}_z \iint J_z \, F(v_1, v_2) \, dv_1 \, dv_2. \end{aligned} \tag{3.79}$$

The three scalar integrals in (3.79) are of Type I with different integrands.

An integral of Type IV is equivalent to

$$I_4 = \iint (J_x\, F_x + J_y\, F_y + J_z\, F_z)\, dS \tag{3.80}$$

which belongs to Type I. Here we have omitted the functional dependence of these functions and the Jacobians on (v_1, v_2).

Finally, an integral of Type V is equivalent to

$$\begin{aligned}\mathbf{I}_5 &= \mathbf{u}_x \iint (J_z\, F_y - J_y\, F_z)\, dv_1\, dv_2 \\ &+ \mathbf{u}_y \iint (J_x\, F_z - J_z\, F_x)\, dv_1\, dv_2 \\ &+ \mathbf{u}_z \iint (J_y\, F_x - J_x\, F_y)\, dv_1\, dv_2.\end{aligned} \tag{3.81}$$

All of the scalar integrals in (3.81) are of Type I. In essence, an integral of Type I is the basic one; all other types of integrals can be reduced to that type. The choice of (v_1, v_2) depends greatly upon the exact nature of the problem. Many integrals resulting from the formulation of physical problems may not be evaluated in a closed form. In these cases, we can seek the help of a numerical method.

3-4 CLASSIFICATION OF VOLUME INTEGRALS

There are only two types of volume integrals:

Type I:

$$I_1 = \iiint_V F(\mathbf{r}_p)\, dV \tag{3.82}$$

Type II:

$$\mathbf{I}_2 = \iiint_V \mathbf{F}(\mathbf{r}_p)\, dV \tag{3.83}$$

where V denotes the domain of integration, which can be either bounded or covering the entire space. We now have three independent variables. In an orthogonal system, they are (v_1, v_2, v_3). An integral of Type I, when expressed in that system, becomes

$$I_1 = \iiint_V F(v_1, v_2, v_3)\, h_1\, h_2\, h_3\, dv_1\, dv_2\, dv_3. \tag{3.84}$$

The choice of the proper coordinate system depends greatly upon the shape of V. From the point of view of the numerical method, we can always use a Cartesian coordinate system to partition the region of integration.

An integral of Type II is equivalent to

$$\mathbf{I}_2 = \mathbf{u}_x \iiint F_x \, dV + \mathbf{u}_y \iiint F_y \, dV + \mathbf{u}_z \iiint F_z \, dV. \tag{3.85}$$

The three scalar integrals in (3.85) are of Type I. We will not discuss the actual execution of evaluating (3.84). The method is found in many books on calculus.

CHAPTER 4

Vector Analysis in Space

4-1 SYMBOLIC VECTOR AND SYMBOLIC VECTOR EXPRESSIONS

This is the most important chapter in the entire book. A new method, designated the *symbolic vector method*, in treating vector analysis will be introduced. The main advantages of this method are

1. the differential expressions of the three key functions in vector analysis are derived based on one basic formula,

2. all of the integral theorems in vector analysis are deduced from one generalized theorem,

3. the commonly used vector identities are found by an algebraic method without performing any differentiation, and

4. two differential operators in the curvilinear coordinate system are introduced. They are designated as the divergence operator and the curl operator, which are distinct from the operator used for the gradient. The technical meaning of the words "divergence," "curl", and "gradient" will be explained shortly. It should be pointed out that the nomenclature for some technical terms introduced in this chapter is different from the original one used in [18].

Since vector algebra is the germ of the method, we will review several essential topics covered in Chapter 1. In vector algebra, there are various products, such as

$$a\mathbf{b}, \ \mathbf{a}\cdot\mathbf{b}, \ \mathbf{a}\times\mathbf{b}, \ c(\mathbf{a}\cdot\mathbf{b}), \ c(\mathbf{a}\times\mathbf{b}), \ \mathbf{c}\cdot(\mathbf{a}\times\mathbf{b}), \ \mathbf{c}\times(\mathbf{a}\times\mathbf{b}),$$
$$(\mathbf{a}\times\mathbf{b})\cdot(\mathbf{c}\times\mathbf{d}). \tag{4.1}$$

All of them have well-defined meanings in vector algebra. Here, we treat the scalar and vector quantities a, $\mathbf{a}$, b, $\mathbf{b}$, c, $\mathbf{c}$, $\mathbf{d}$ as functions of position, and they are assumed to be distinct from each other. For the purpose of identification, the functions listed in (4.1) will be referred to as *vector expressions*. A quantity like $\mathbf{ab}$ is not a vector expression, although it is a well-defined quantity in dyadic analysis, a subject to be treated in a later chapter. For the time being, we are dealing with vector expressions only. In one case, a dyadic quantity will be involved. Its implication will be explained then. All of the vector expressions listed in (4.1) are linear with respect to a single function, i.e., the distributive law holds true. For example, if $\mathbf{c} = \mathbf{c}_1 + \mathbf{c}_2$, then

$$\mathbf{c} \cdot (\mathbf{a} \times \mathbf{b}) = (\mathbf{c}_1 + \mathbf{c}_2) \cdot (\mathbf{a} \times \mathbf{b}) = \mathbf{c}_1 \cdot (\mathbf{a} \times \mathbf{b}) + \mathbf{c}_2 \cdot (\mathbf{a} \times \mathbf{b}). \tag{4.2}$$

There are several important identities in vector algebra which are listed below:

$$a\mathbf{b} = \mathbf{b}a \tag{4.3}$$

$$\mathbf{a} \cdot \mathbf{b} = \mathbf{b} \cdot \mathbf{a} \tag{4.4}$$

$$\mathbf{a} \times \mathbf{b} = -\mathbf{b} \times \mathbf{a} \tag{4.5}$$

$$\mathbf{c} \cdot (\mathbf{a} \times \mathbf{b}) = \mathbf{b} \cdot (\mathbf{c} \times \mathbf{a}) = \mathbf{a} \cdot (\mathbf{b} \times \mathbf{c}) \tag{4.6}$$

$$\mathbf{c} \times (\mathbf{a} \times \mathbf{b}) = (\mathbf{c} \cdot \mathbf{b})\mathbf{a} - (\mathbf{c} \cdot \mathbf{a})\mathbf{b}. \tag{4.7}$$

The proofs of (4.6)–(4.7) are found in section 2 of Chapter 1.

Now, if one of the vectors in (4.1) is replaced by a symbolic vector denoted by ∇ such as $a\nabla$, $\nabla \cdot \mathbf{b}$, $\mathbf{a} \cdot \nabla$, $\nabla \times \mathbf{b}$, $\mathbf{a} \times \nabla$, $c\nabla \cdot \mathbf{b}$, $c\mathbf{a} \cdot \nabla$, $\mathbf{c} \cdot (\nabla \times \mathbf{b})$, etc., these expressions will be called *symbolic vector expressions* or *symbolic expressions* for short. "∇" is designated as the *symbolic vector* or *S vector* for short. Besides ∇, a symbolic expression contains other functions, either scalar or/and vector. Thus, $c(\nabla \times \mathbf{b})$ contains one scalar, one vector, and the S vector. In general, a symbolic expression will be denoted by $T(\nabla)$ or, more specifically, by $T(\nabla, f_1, f_2, ..., f_N)$ if there is a need to identify the functions contained in the expression. These functions can be either scalar, vector, or a combination as long as it has an acceptable form in vector algebra, i.e., it is created from a valid vector expression by replacing one of the vectors by the S vector ∇. The symbolic expression so created is *defined* by

$$T(\nabla) = \lim_{\Delta V \to 0} \frac{\sum_i T(\mathbf{n}_i)\Delta S_i}{\Delta V} \tag{4.8}$$

where ΔS_i denotes a typical elementary area of a surface enclosing the volume ΔV of a small cell, and $\mathbf{n}_i$ denotes the unit outward normal vector from ΔS_i. The running index i in (4.8) covers the number of surfaces enclosing ΔV. For a cubical volume, it goes from 1 to 6. Since this definition is independent of the choice of the coordinate system or *invariant to the coordinate system*, we can find the differential expression of $T(\nabla)$ in any coordinate system by evaluating the limit value of the right-hand side term of (4.8). For the general curvilinear orthogonal system with coordinate variables (v_1, v_2, v_3), unit vectors $(\mathbf{u}_1, \mathbf{u}_2, \mathbf{u}_3)$, and metric coefficients (h_1, h_2, h_3), the differential expression of $T(\nabla)$ can be found by considering surfaces located at $v_i \pm \Delta v_i/2$ with $i = 1, 2, 3$, as shown in Fig. 4-1. The contributions from various surfaces to the sum in the numerator are

$$\sum_{i=1}^{6} T(\mathbf{n}_i)\Delta S_i = T(\mathbf{u}_1)h_2h_3\Delta v_2\Delta v_3|_{v_1+\frac{\Delta v_1}{2}} + T(-\mathbf{u}_1)h_2h_3\Delta v_2\Delta v_3|_{v_1-\frac{\Delta v_1}{2}}$$
$$+T(\mathbf{u}_2)h_1h_3\Delta v_1\Delta v_3|_{v_2+\frac{\Delta v_2}{2}} + T(-\mathbf{u}_2)h_1h_3\Delta v_1\Delta v_3|_{v_2-\frac{\Delta v_2}{2}}$$
$$+T(\mathbf{u}_3)h_1h_2\Delta v_1\Delta v_2|_{v_3+\frac{\Delta v_3}{2}} + T(-\mathbf{u}_3)h_1h_2\Delta v_1\Delta v_2|_{v_3-\frac{\Delta v_3}{2}}. \quad (4.9)$$

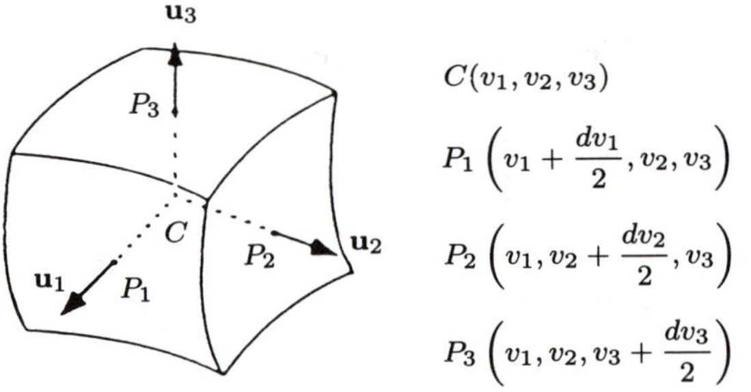

Fig. 4-1 Cell bounded by six coordinate surfaces in a curvilinear orthogonal system with the center at C.

Since $T(\nabla)$ is linear with respect to ∇, because the symbolic expression is generated from a vector expression with one of vectors replaced by ∇, $T(\mathbf{n}_i)$ is linear with respect to $\mathbf{n}_i$; thus, $T(-\mathbf{u}_i) = -T(\mathbf{u}_i)$. Upon substituting (4.9) into (4.8) with $\Delta V = h_1h_2h_3\Delta v_1\Delta v_2\Delta v_3$, and taking the limit, we find

$$T(\nabla) = \frac{1}{h_1h_2h_3}\left\{\frac{\partial}{\partial v_1}[h_2h_3T(\mathbf{u}_1)] + \frac{\partial}{\partial v_2}[h_1h_3T(\mathbf{u}_2)] + \frac{\partial}{\partial v_3}[h_1h_2T(\mathbf{u}_3)]\right\}. \quad (4.10)$$

It is implied here that the metric coefficients, the unit vectors, and the functions contained in $T(\mathbf{u}_i)$ with $i = 1, 2, 3$ are, in general, functions of position. A compact formula for (4.10) is

$$T(\nabla) = \frac{1}{\Omega}\sum_i \frac{\partial}{\partial v_i}\left[\frac{\Omega}{h_i}T(\mathbf{u}_i)\right] \quad (4.11)$$

where $\Omega = h_1h_2h_3$ and the running index i goes from 1 to 3. This labeling will be omitted unless stated otherwise. In a Cartesian system, (4.11) reduces to

$$T(\nabla) = \sum_i \frac{\partial}{\partial x_i}T(\mathbf{a}_i) \quad (4.12)$$

where we have used (x_1, x_2, x_3) for (x, y, z) and $(\mathbf{a}_1, \mathbf{a}_2, \mathbf{a}_3)$for $(\mathbf{u}_x, \mathbf{u}_y, \mathbf{u}_z)$ in order to use the summation sign. This notation will be adopted in the whole book whenever there is no confusion. The invariant property of $T(\nabla)$ means that if (4.11) is evaluated in any system, say the cylindrical coordinate system, the resultant differential expression is functionally equal to another one evaluated in a different system such as (4.12) which is expressed in the Cartesian system.

Before we proceed further, an important lemma or rule, to be used frequently in the following sections, will be introduced. The statement of this lemma is as follows.

Lemma 1. For any symbolic expression $T(\nabla)$, which is generated from a valid vector expression, we can treat the S vector ∇ in that expression as a vector, and all of the algebraic identities in vector algebra are applicable.

For example, we have some important vector identities listed in (4.3)–(4.7); then, according to Lemma 1, the following relations hold true:

$$a\nabla = \nabla a \tag{4.13}$$

$$\nabla \cdot \mathbf{a} = \mathbf{a} \cdot \nabla \tag{4.14}$$

$$\nabla \times \mathbf{a} = -\mathbf{a} \times \nabla \tag{4.15}$$

$$\mathbf{b} \cdot (\mathbf{a} \times \nabla) = \nabla \cdot (\mathbf{b} \times \mathbf{a}) = \mathbf{a} \cdot (\nabla \times \mathbf{b}) \tag{4.16}$$

$$\nabla \times (\mathbf{a} \times \mathbf{b}) = (\nabla \cdot \mathbf{b})\,\mathbf{a} - (\nabla \cdot \mathbf{a})\,\mathbf{b}. \tag{4.17}$$

The proof of Lemma 1 follows directly from the definition of $T(\nabla)$ stated by (4.8) and its differential expression by (4.11) because if we replace the S vector in (4.13)–(4.17) by $\mathbf{n}_i$ or $\mathbf{u}_i$, they are valid identities in vector algebra, of exactly the same form as (4.3)–(4.7).

4-2 GRADIENT, DIVERGENCE, AND CURL

Let us now consider some simple symbolic expressions which are generated by using vector expressions with two functions, one of which must be a vector. There are only three possibilities, namely, $\mathbf{a}f$, $\mathbf{a} \cdot \mathbf{f}$, and $\mathbf{a} \times \mathbf{f}$. By replacing $\mathbf{a}$ with ∇, we generate three symbolic expressions with the form ∇f, $\nabla \cdot \mathbf{f}$, and $\nabla \times \mathbf{f}$. According to Lemma 1, we have

$$\nabla f = f\nabla\,, \quad \nabla \cdot \mathbf{f} = \mathbf{f} \cdot \nabla\,, \quad \nabla \times \mathbf{f} = -\mathbf{f} \times \nabla. \tag{4.18}$$

This is an important characteristic of these symbolic expressions. Let us now reveal the differential forms of these expressions using (4.12), stated in the Cartesian system, which is repeated here:

$$T(\nabla) = \sum_i \frac{\partial}{\partial x_i} T(\mathbf{a}_i). \tag{4.19}$$

1. Gradient

   When $T(\nabla) = \nabla f$, $T(\mathbf{a}_i) = f\mathbf{a}_i$; upon substituting it into (4.12), we obtain

$$\nabla f = \sum_i \frac{\partial}{\partial x_i}(f\mathbf{a}_i) = \sum_i \mathbf{a}_i \frac{\partial f}{\partial x_i}. \tag{4.20}$$

The differential function so obtained is designated as the gradient of a scalar function or the *gradient* for short.

2. Divergence

When $T(\nabla) = \nabla \cdot \mathbf{f}$, $T(\mathbf{a}_i) = \mathbf{a}_i \cdot \mathbf{f}$; upon substituting it into (4.12), we obtain

$$\nabla \cdot \mathbf{f} = \sum_i \frac{\partial}{\partial x_i}(\mathbf{a}_i \cdot \mathbf{f}) = \sum_i \frac{\partial f_i}{\partial x_i}. \tag{4.21}$$

The differential function is designated as the divergence of a vector function $\mathbf{f}$ or the *divergence* for short.

3. Curl

When $T(\nabla) = \nabla \times \mathbf{f}$, $T(\mathbf{a}_i) = \mathbf{a}_i \times \mathbf{f} = -\mathbf{f} \times \mathbf{a}_i$; we have

$$\begin{aligned}\nabla \times \mathbf{f} &= \sum_i \frac{\partial}{\partial x_i}(\mathbf{a}_i \times \mathbf{f}) = \sum_i \mathbf{a}_i \times \frac{\partial \mathbf{f}}{\partial x_i}\\ &= \sum_i \mathbf{a}_i \times \frac{\partial}{\partial x_i}\sum_j f_j \mathbf{a}_j\\ &= \left(\frac{\partial f_3}{\partial x_2} - \frac{\partial f_2}{\partial x_3}\right)\mathbf{a}_1 + \left(\frac{\partial f_1}{\partial x_3} - \frac{\partial f_3}{\partial x_1}\right)\mathbf{a}_2 + \left(\frac{\partial f_2}{\partial x_1} - \frac{\partial f_1}{\partial x_2}\right)\mathbf{a}_3.\end{aligned} \tag{4.22}$$

The function so derived is designated as the curl of a vector function $\mathbf{f}$ or the *curl* for short.

All of these functions are so far defined in the Cartesian system. The form of the gradient suggests that one can define a differential operator, denoted by ∇, in the form

$$\nabla = \sum_i \mathbf{a}_i \frac{\partial}{\partial x_i} = \mathbf{u}_x \frac{\partial}{\partial x} + \mathbf{u}_y \frac{\partial}{\partial y} + \mathbf{u}_z \frac{\partial}{\partial z} \tag{4.23}$$

which is called the del operator (an inverted Greek letter), the Nabla operator (the shape of a lute), or the Hamilton operator, in honor of the English scientist William Rowen Hamilton (1806–1865) who first introduced it, except that his original symbol is " $\triangleright$," not "∇." When this operator is applied to a scalar function f, we obtain

$$\left(\sum_i \mathbf{a}_i \frac{\partial}{\partial x_i}\right) f = \sum_i \mathbf{a}_i \frac{\partial f}{\partial x_i} \tag{4.24}$$

where we have enforced the distributive rule on such an operation.

There are two sets of notations for these three key functions in vector analysis. The older notations are grad f, div $\mathbf{f}$, and curl $\mathbf{f}$ or rot $\mathbf{f}$, which have been used by many European authors. The modern notations for these functions were introduced by J. Willard Gibbs (1839–1903), a renowned American physicist, over a century ago. They are ∇f for the gradient, $\nabla \cdot \mathbf{f}$ for the divergence, and $\nabla \times \mathbf{f}$ for the curl. It is seen that the del operator in ∇f is truly a differential operator defined by (4.23), but the symbol for the del operator in $\nabla \cdot \mathbf{f}$ and $\nabla \times \mathbf{f}$ does not have the same significance as the one in ∇f. One must not interpret $\nabla \cdot \mathbf{f}$ and $\nabla \times \mathbf{f}$ as the "scalar product" and the "vector product" between ∇ and $\mathbf{f}$. In fact, these products do not exist. We shall elaborate on this remark later, after we have derived the differential expressions for these functions in the general curvilinear orthogonal system that includes the Cartesian system as a special case. In summary, the old and modern notations and the differential expressions for the three key functions in vector analysis in a Cartesian system are

$$\text{grad } f = \nabla f = \sum_i \mathbf{a}_i \frac{\partial f}{\partial x_i} \tag{4.25}$$

$$\text{div } \mathbf{f} = \nabla \cdot \mathbf{f} = \sum_i \frac{\partial f_i}{\partial x_i} \tag{4.26}$$

$$\text{curl } \mathbf{f} = \nabla \times f = \sum_i \left(\frac{\partial f_k}{\partial x_j} - \frac{\partial f_j}{\partial x_k} \right) \mathbf{a}_i. \tag{4.27}$$

In (4.27), $(i, j, k) = (1, 2, 3)$ in cyclic order. Historically, these are the expressions used by Gibbs [5] to *define* these functions in the Cartesian system.

In this book, only the modern notations for these functions will be used in the subsequent sections and chapters. We repeat once more that no meaning should be given to the modern notations for the divergence and the curl. They are merely *notations*! Since

$$\frac{\partial \mathbf{f}}{\partial x_i} = \frac{\partial}{\partial x_i} \sum_j f_j \mathbf{a}_j = \sum_j \frac{\partial f_j}{\partial v_i} \mathbf{a}_j$$

because $\mathbf{a}_j$ is a constant vector, and furthermore, we have the relations

$$\mathbf{a}_i \cdot \mathbf{a}_j = \begin{cases} 1, & i = j \\ 0, & i \neq j \end{cases}$$

$$\mathbf{a}_i = \mathbf{a}_j \times \mathbf{a}_k$$

with $(i, j, k) = (1, 2, 3)$ in cyclic order, then (4.26) and (4.27) can also be written in the form

$$\nabla \cdot \mathbf{f} = \sum_i \mathbf{a}_i \cdot \frac{\partial \mathbf{f}}{\partial x_i} \tag{4.28}$$

$$\nabla \times \mathbf{f} = \sum_i \mathbf{a}_i \times \frac{\partial \mathbf{f}}{\partial x_i}. \tag{4.29}$$

Thus, in the Cartesian system, the differential operator for the divergence has the form

$$\sum_i \mathbf{a}_i \cdot \frac{\partial}{\partial x_i}, \tag{4.30}$$

and the differential operator for the curl has the form

$$\sum_i \mathbf{a}_i \times \frac{\partial}{\partial x_i}. \tag{4.31}$$

They are distinctly different from the del operator used in ∇f as defined by (4.23). When we compare (4.20)–(4.22) to (4.25)–(4.27), it appears that the S vector $\forall$ is identical to the del operator ∇. But this is not true; there is a fundamental difference between them. Thus, according to (4.18),

$$\forall f = f \forall,$$

but

$$\nabla f \neq f \nabla \tag{4.32}$$

because $f\nabla$ is a weighted differential operator, not a function. It should be emphasized that $\forall$ is a symbolic vector; *it is not a differential operator, and* $\nabla \cdot \mathbf{f}$ *and* $\nabla \times \mathbf{f}$ *are merely notations for the divergence and the curl.* In other words, Lemma 1 applies to $\forall f$ or $T(\forall)$ in general, but not to ∇f. In tracing back to the original definition of $T(\forall)$ given by (4.8), we see that three key functions in vector analysis have the general definition of

$$\nabla f = \lim_{\Delta V \to 0} \frac{\sum_i \mathbf{n}_i f \Delta S_i}{\Delta V} \tag{4.33}$$

$$\nabla \cdot \mathbf{f} = \lim_{\Delta V \to 0} \frac{\sum_i \mathbf{n}_i \cdot \mathbf{f} \Delta S_i}{\Delta V} \tag{4.34}$$

$$\nabla \times \mathbf{f} = \lim_{\Delta V \to 0} \frac{\sum_i \mathbf{n}_i \times \mathbf{f} \Delta S_i}{\Delta V}. \tag{4.35}$$

To distinguish these functions in a conceptual manner, we propose some names for the quantities involved in (4.33)–(4.35) based on a "physical" model. Thus, the term $\sum \mathbf{n}_i f \Delta S_i$ in (4.33), which is a vector quantity, will be identified as the total directional radiance of f from the volume cell ΔV or *radiance* for short; the gradient is then a measure of radiance per unit volume. The term $\sum \mathbf{n}_i \cdot \mathbf{f} \Delta S_i$ in (4.34) is a scalar quantity. It has a well-established name used by many engineers and scientists as the total flux of $\mathbf{f}$ from ΔV or *flux* for short; the divergence is then a measure of flux per unit volume. For the vector quantity $\sum \mathbf{n}_i \times \mathbf{f} \Delta S_i$ in (4.35), we propose the name of the total shear of $\mathbf{f}$ around the enclosing volume ΔV or *shear* for short; the curl is then a measure of shear per unit volume. From the mathematical point of view, there is no need to inject this "physical" model. It is proposed here, particularly for the gradient and the curl, merely as an aid to distinguish these functions based on a more practical model.

After showing the differential form of the three key functions in a Cartesian system, we are going to derive their differential expressions in the general curvilinear coordinate system based on (4.11), which is repeated here:

$$T(\nabla) = \frac{1}{\Omega} \sum_i \frac{\partial}{\partial v_i} \left[\frac{\Omega}{h_i} T(\mathbf{u}_i) \right]. \tag{4.36}$$

Consider first the symbolic expression for the gradient, i.e., $T(\nabla) = \nabla f = \nabla f$; then $T(\mathbf{u}_i) = f\mathbf{u}_i$; hence,

$$\begin{aligned} \nabla f &= \frac{1}{\Omega} \sum_i \frac{\partial}{\partial v_i} \left[\frac{\Omega f}{h_i} \mathbf{u}_i \right] \\ &= \frac{1}{\Omega} \sum_i \left[\frac{\partial}{\partial v_i} \left(\frac{\Omega \mathbf{u}_i}{h_i} \right) f + \frac{\Omega \mathbf{u}_i}{h_i} \frac{\partial f}{\partial v_i} \right]. \end{aligned} \tag{4.37}$$

The coefficient associated with f in (4.37) vanishes because of (2.27); thus,

$$\nabla f = \sum_i \frac{\mathbf{u}_i}{h_i} \frac{\partial f}{\partial v_i}. \tag{4.38}$$

For the divergence, $T(\nabla) = \nabla \cdot \mathbf{f} = \nabla \cdot \mathbf{f}$, and $T(\mathbf{u}_i) = \mathbf{u}_i \cdot \mathbf{f} = f_i$; hence,

$$\begin{aligned} \nabla \cdot \mathbf{f} &= \frac{1}{\Omega} \sum_i \frac{\partial}{\partial v_i} \left[\frac{\Omega}{h_i} \mathbf{u}_i \cdot \mathbf{f} \right] \\ &= \frac{1}{\Omega} \sum_i \frac{\partial}{\partial v_i} \left(\frac{\Omega}{h_i} f_i \right). \end{aligned} \tag{4.39}$$

For the curl, $T(\nabla) = \nabla \times \mathbf{f} = \nabla \times \mathbf{f}$, so $T(\mathbf{u}_i) = \mathbf{u}_i \times \mathbf{f}$; then

$$\begin{aligned} \nabla \times \mathbf{f} &= \frac{1}{\Omega} \sum_i \frac{\partial}{\partial v_i} \left[\frac{\Omega}{h_i} \mathbf{u}_i \times \mathbf{f} \right] \\ &= \frac{1}{\Omega} \left[\frac{\partial}{\partial v_1} (h_2 h_3 \mathbf{u}_1 \times \mathbf{f}) + \frac{\partial}{\partial v_2} (h_1 h_3 \mathbf{u}_2 \times \mathbf{f}) + \frac{\partial}{\partial v_3} (h_1 h_2 \mathbf{u}_3 \times \mathbf{f}) \right] \\ &= \frac{1}{\Omega} \left[\frac{\partial}{\partial v_1} h_2 h_3 (f_2 \mathbf{u}_3 - f_3 \mathbf{u}_2) \right. \\ &\quad + \frac{\partial}{\partial v_2} h_1 h_3 (-f_1 \mathbf{u}_3 + f_3 \mathbf{u}_1) \\ &\quad \left. + \frac{\partial}{\partial v_3} h_1 h_2 (f_1 \mathbf{u}_2 - f_2 \mathbf{u}_1) \right]. \end{aligned} \tag{4.40}$$

A term like $1/\Omega\ \partial/\partial v_1 (h_2 h_3 f_2 \mathbf{u}_3)$ can be split into two terms:

$$\frac{1}{\Omega} \frac{\partial}{\partial v_1} (h_2 h_3 f_2 \mathbf{u}_3) = \frac{1}{\Omega} \left[h_2 f_2 \frac{\partial (h_3 \mathbf{u}_3)}{\partial v_1} + h_3 \mathbf{u}_3 \frac{\partial (h_2 f_2)}{\partial v_1} \right],$$

and similarly, for the remaining five terms, which are all of the form

$$\frac{1}{\Omega}\frac{\partial}{\partial v_i}(h_j h_k f_j \mathbf{u}_k), \qquad (i \neq j \neq k).$$

As a result of (2.21), we find that terms involving the derivatives of $\partial(h_i\mathbf{u}_i)/\partial v_j$ (with $i, j = 1, 2, 3$, and $i \neq j$) cancel each other. The remaining terms can be arranged in a "determinant" form:

$$\nabla \times \mathbf{f} = \frac{1}{\Omega}\begin{vmatrix} h_1\mathbf{u}_1 & h_2\mathbf{u}_2 & h_3\mathbf{u}_3 \\ \dfrac{\partial}{\partial v_1} & \dfrac{\partial}{\partial v_2} & \dfrac{\partial}{\partial v_3} \\ h_1 f_1 & h_2 f_2 & h_3 f_3 \end{vmatrix}. \tag{4.41}$$

In the special case when $(v_1, v_2, v_3) = (x_1, x_2, x_3)$, $(\mathbf{u}_1, \mathbf{u}_2, \mathbf{u}_3) = (\mathbf{a}_1, \mathbf{a}_2, \mathbf{a}_3)$, and $h_i = 1, i = 1, 2, 3$, (4.38), (4.39), and (4.41) reduce to the differential form of these functions in a Cartesian system. Because of the invariance principle, these functions are equal to each other no matter what coordinate system is used to express their differential form. It may be of interest to summarize the previous formulas for the three basic functions in one compact abstract format by defining

$$T(\nabla) = \nabla * \tilde{g} \tag{4.42}$$

where the meaning of the asterisk "$*$" and the function g with the tilde is explained in Table 4-1.

TABLE 4-1
Abstract Format

$*$	$\tilde{g}$	$\nabla * \tilde{g} = \nabla * \tilde{g}$	Name of the Function
Null	f	$\nabla f = \nabla f$	Gradient
$\cdot$	$\mathbf{f}$	$\nabla \cdot \mathbf{f} = \nabla \cdot \mathbf{f}$	Divergence
$\times$	$\mathbf{f}$	$\nabla \times \mathbf{f} = \nabla \times \mathbf{f}$	Curl

In terms of this format, we can write

$$\begin{aligned} \nabla * \tilde{g} &= \lim_{\Delta V \to 0} \frac{\sum_i \mathbf{n}_i * \tilde{g}\Delta S_i}{\Delta V} \\ &= \frac{1}{\Omega}\sum_i \frac{\partial}{\partial v_i}\left[\frac{\Omega}{h_i}\mathbf{u}_i * \tilde{g}\right] \\ &= \frac{1}{\Omega}\sum_i \left[\frac{\partial}{\partial v_i}\left(\frac{\Omega}{h_i}\mathbf{u}_i\right) * \tilde{g} + \frac{\Omega}{h_i}\mathbf{u}_i * \frac{\partial \tilde{g}}{\partial v_i}\right] \end{aligned} \tag{4.43}$$

$$= \sum_i \frac{1}{h_i} \mathbf{u}_i * \frac{\partial \tilde{g}}{\partial v_i}$$

$$= \nabla * \tilde{g} \tag{4.44}$$

where

$$\nabla = \sum_i \frac{\mathbf{u}_i}{h_i} \frac{\partial}{\partial v_i} \tag{4.45}$$

is the del operator defined in the general curvilinear coordinate system. The format $\nabla * \tilde{g}$ as defined by (4.44) was used by this author [16] as a unified definition for the gradient, the divergence, and the curl before the symbolic method was found, and it was extended to dyadic functions by Fang and Zhu [3]. The definitions of gradient, divergence, and curl in the form of (4.33)–(4.35) have previously been discussed by Gans [4]. In our derivation of the differential expressions of these functions in the general orthogonal system, we have made full use of various relationships of the derivatives of the unit vectors. The symbolic method makes the presentation even more systematic.

In view of (4.44)–(4.45), the expressions for the gradient, divergence, and curl in the general orthogonal system can also be written in the form

$$\nabla f = \sum_i \frac{\mathbf{u}_i}{h_i} \frac{\partial f}{\partial v_i} \tag{4.46}$$

$$\nabla \cdot \mathbf{f} = \sum_i \frac{\mathbf{u}_i}{h_i} \cdot \frac{\partial \mathbf{f}}{\partial v_i} \tag{4.47}$$

$$\nabla \times \mathbf{f} = \sum_i \frac{\mathbf{u}_i}{h_i} \times \frac{\partial \mathbf{f}}{\partial v_i}. \tag{4.48}$$

It is seen that there are three distinct operators involved. For the gradient, we have the ordinary del operator. For the divergence, we have a differential operator which will be denoted by $\nabla\!\!\!\cdot\,$ and designated as the divergence operator or the dot–del operator. It is defined by

$$\nabla\!\!\!\cdot\, = \sum_i \frac{\mathbf{u}_i}{h_i} \cdot \frac{\partial}{\partial v_i}. \tag{4.49}$$

When this operator is applied to a vector function, it yields the differential expression for the divergence in the general curvilinear orthogonal system, i.e.,

$$\nabla\!\!\!\cdot\, \mathbf{f} = \nabla \cdot \mathbf{f}. \tag{4.50}$$

For the curl, we have another differential operator, which will be denoted by $\nabla\!\!\!\!\times$ and designated as the curl operator or the cross–del operator. It is defined by

$$\nabla\!\!\!\!\times = \sum_i \frac{\mathbf{u}_i}{h_i} \times \frac{\partial}{\partial v_i}. \tag{4.51}$$

When this operator is applied to a vector function, the result gives the differential expression for the curl in the general curvilinear orthogonal system, i.e.,

$$\overset{\times}{\nabla}\mathbf{f} = \nabla \times \mathbf{f}. \tag{4.52}$$

It is understood that we have enforced the distributive rule on these operations. For the Cartesian system, $h_i = 1, v_i = x_i, \mathbf{u}_i = \mathbf{a}_i$, we recover (4.25), (4.28), and (4.29). These new notations for the divergence and the curl are quite descriptive. However, we are not suggesting that they should be used to replace the long-established notations introduced by Gibbs. For classroom work, these notations may be more convenient. At least, they would certainly eliminate the possibility of "interpreting" $\nabla \cdot \mathbf{f}$ and $\nabla \times \mathbf{f}$ as the "scalar product" and the "vector product" between ∇ and $\mathbf{f}$. The confusion resulting from this kind of false interpretation of Gibbs' notations has recently been pointed out by this author. The work is reproduced in Appendix E as a reminder of a historical episode. Some of the material is also covered in the text.

Finally, it should be mentioned that in many books on applied mathematics, the invariance principle is usually demonstrated by starting with the definition of the three basic functions defined in a Cartesian system, and then deriving their differential expressions in other coordinate systems by a functional transformation. For example, in the case of divergence, its differential expression in the Cartesian system is

$$\nabla \cdot \mathbf{f} = \frac{\partial f_x}{\partial x} + \frac{\partial f_y}{\partial y} + \frac{\partial f_z}{\partial z}.$$

To find its expression, say, in the cylindrical coordinate system, we use the relations

$$\begin{aligned} \mathbf{f} &= f_x \mathbf{u}_x + f_y \mathbf{u}_y + f_z \mathbf{u}_z \\ &= f_r \mathbf{u}_r + f_\phi \mathbf{u}_\phi + f_z \mathbf{u}_z \end{aligned}$$

so

$$\begin{aligned} \mathbf{u}_x \cdot \mathbf{f} &= f_x = (\mathbf{u}_x \cdot \mathbf{u}_r) f_r + (\mathbf{u}_x \cdot \mathbf{u}_\phi) f_\phi \\ &= \cos\phi f_r - \sin\phi f_\phi. \end{aligned}$$

Similarly,

$$\mathbf{u}_y \cdot \mathbf{f} = f_y = \sin\phi f_r + \cos\phi f_\phi.$$

In a later section, this kind of relationship will be derived for more complicated coordinate systems. The relationships between the coordinate variables are

$$x = r\cos\phi, \quad y = r\sin\phi, \quad z = z;$$

hence,

$$\frac{\partial r}{\partial x} = \cos\phi, \quad \frac{\partial r}{\partial y} = \sin\phi,$$

$$\frac{\partial \phi}{\partial x} = -\frac{\sin\phi}{r}, \quad \frac{\partial \phi}{\partial y} = \frac{\cos\phi}{r}.$$

By the rule of chain differentiation, we obtain

$$\begin{aligned}\frac{\partial f_x}{\partial x} &= \frac{\partial}{\partial x}(\cos\phi f_r - \sin\phi f_\phi) \\ &= \frac{\partial}{\partial r}(\cos\phi f_r - \sin\phi f_\phi)\frac{\partial r}{\partial x} + \frac{\partial}{\partial \phi}(\cos\phi f_r - \sin\phi f_\phi)\frac{\partial \phi}{\partial x} \\ &= \left(\cos\phi\frac{\partial f_r}{\partial r} - \sin\phi\frac{\partial f_\phi}{\partial r}\right)\cos\phi \\ &\quad + \left(\cos\phi\frac{\partial f_r}{\partial \phi} - \sin\phi\frac{\partial f_\phi}{\partial \phi} - \sin\phi f_r - \cos\phi f_\phi\right)\frac{-\sin\phi}{r}.\end{aligned}$$

Similarly,

$$\begin{aligned}\frac{\partial f_y}{\partial y} &= \left(\sin\phi\frac{\partial f_r}{\partial r} + \cos\phi\frac{\partial f_\phi}{\partial r}\right)\sin\phi \\ &\quad + \left(\sin\phi\frac{\partial f_r}{\partial \phi} + \cos\phi\frac{\partial f_\phi}{\partial \phi} + \cos\phi f_r - \sin\phi f_\phi\right)\frac{\cos\phi}{r}.\end{aligned}$$

The sum of these two terms, together with $\partial f_z/\partial z$, yields

$$\begin{aligned}\nabla\cdot\mathbf{f} &= \frac{\partial f_r}{\partial r} + \frac{f_r}{r} + \frac{1}{r}\frac{\partial f_\phi}{\partial \phi} + \frac{\partial f_z}{\partial z} \\ &= \frac{1}{r}\frac{\partial}{\partial r}(r f_r) + \frac{1}{r}\frac{\partial f_\phi}{\partial \phi} + \frac{\partial f_z}{\partial z},\end{aligned}$$

which is identical to (4.39) with $(v_1, v_2, v_3) = (r, \phi, z)$, and $(h_1, h_2, h_3) = (1, r, 1)$. The functional transformation of the gradient and the curl can be carried out in a similar manner. In these cases, the unit vectors are also involved. It is evident that the relationships between the unit vectors of various systems and the ones in the Cartesian system must be derived first before the functional transformation can be executed. This example also demonstrated very clearly the advantage of using (4.8), and hence (4.11), to derive the differential expression of these basic functions in any orthogonal coordinate system.

4-3 ALTERNATIVE DEFINITION OF GRADIENT AND CURL

In the defining equation for ∇f as given by (4.33), let the shape of ΔV be a flat cell of uniform thickness Δs and area ΔA at the broad surfaces, as shown in Fig. 4-2. The outward unit normal vector is denoted by $\mathbf{n}_i$. By taking a scalar product of that equation with $\mathbf{u}_s$, we obtain

$$\begin{aligned}\mathbf{u}_s\cdot\nabla f &= \lim_{\Delta V\to 0}\frac{\sum_i f\,(\mathbf{u}_s\cdot\mathbf{n}_i)\,\Delta S_i}{\Delta V} \\ &= \lim_{\Delta A\to 0, \Delta s\to 0}\frac{\sum_i f\,(\mathbf{u}_s\cdot\mathbf{n}_i)\,\Delta S_i}{\Delta A\Delta s}.\end{aligned} \tag{4.53}$$

The scalar product $\mathbf{u}_s \cdot \mathbf{n}_i$ vanishes for all of the side surfaces because $\mathbf{u}_s$ is perpendicular to $\mathbf{n}_i$ therein. The only contribution results from the top and the bottom surfaces where $\mathbf{n}_i = \pm\mathbf{u}_s$ and $\Delta S_i = \Delta A$; thus, we obtain

$$\begin{aligned}\mathbf{u}_s \cdot \nabla f &= \lim_{\Delta s \to 0} \frac{\left[f\left(s + \frac{\Delta s}{2}\right) - f\left(s - \frac{\Delta s}{2}\right)\right]}{\Delta s} \\ &= \frac{\partial f}{\partial s}\end{aligned} \tag{4.54}$$

where $s \pm \Delta s/2$ corresponds to the locations of the two broad surfaces along the s coordinate, and the center of the flat cell is located at s. Equation (4.54), therefore, can be treated as an alternative definition for the scalar component of the gradient in an arbitrary direction $\mathbf{u}_s$.

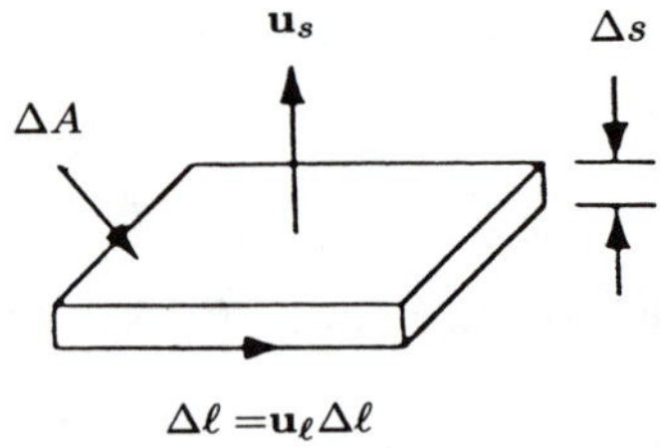

Fig. 4-2 Thin flat volume of uniform thickness Δs and area ΔA.

This result can also be obtained by the rule of chain derivatives. Thus, for any parameter s,

$$\frac{\partial f}{\partial s} = \sum_i \frac{\partial f}{\partial x_i}\frac{\partial x_i}{\partial s}. \tag{4.55}$$

If ds denotes the total differential of the position vector along the s axis, we have

$$ds\ \mathbf{u}_s = \sum_i dx_i \mathbf{a}_i; \tag{4.56}$$

hence,

$$\mathbf{u}_s = \sum_i \frac{\partial x_i}{\partial s}\mathbf{a}_i. \tag{4.57}$$

Equation (4.55), therefore, corresponds to the scalar product of the unit vector $\mathbf{u}_s$ with another vector defined by

$$\sum_i \frac{\partial f}{\partial x_i}\mathbf{a}_i, \tag{4.58}$$

and we designate this vector as the gradient of f, represented by ∇f. In this approach, which is relatively simple, the definition of the scalar component of the gradient in an arbitrary direction is introduced first, and then the full differential expression of the gradient, at least in a Cartesian system, is derived.

To derive the differential expression for the gradient in the general orthogonal system by this approach, we start with

$$\frac{\partial f}{\partial v_i} = \sum_j \frac{\partial f}{\partial x_j}\frac{\partial x_j}{\partial v_i}. \tag{4.59}$$

Now,

$$d\mathbf{s}_i = h_i dv_i \mathbf{u}_i = \sum_j dx_j \mathbf{a}_j; \tag{4.60}$$

hence,

$$\mathbf{u}_i = \frac{1}{h_i}\sum_j \frac{\partial x_j}{\partial v_i}\mathbf{a}_j. \tag{4.61}$$

Equation (4.59) can be changed to

$$\begin{aligned}\frac{1}{h_i}\frac{\partial f}{\partial v_i} &= \frac{1}{h_i}\sum_j \frac{\partial f}{\partial x_j}\frac{\partial x_j}{\partial v_i}\\ &= \left(\frac{1}{h_i}\sum_j \frac{\partial x_j}{\partial v_i}\mathbf{a}_j\right)\cdot\sum_k \frac{\partial f}{\partial x_k}\mathbf{a}_k\\ &= \mathbf{u}_i\cdot\nabla f, \qquad i = 1, 2, 3,\end{aligned} \tag{4.62}$$

which is the scalar component of ∇f in the direction of $\mathbf{u}_i$.

The same model can be used to derive the differential expression for the curl of $\mathbf{f}$. According to (4.35), the general definition of the curl is

$$\begin{aligned}\nabla\times\mathbf{f} &= \lim_{\Delta V\to 0}\frac{\sum_i(\mathbf{n}_i\times\mathbf{f})\,\Delta S_i}{\Delta V}\\ &= \lim_{\Delta V\to 0}\frac{\sum_i(\Delta\mathbf{S}_i\times\mathbf{f})}{\Delta V}.\end{aligned} \tag{4.63}$$

We use the same model of ΔV in Fig. 4-2 for the above expression. By taking a scalar product of (4.63) with $\mathbf{u}_s$, we obtain

$$\begin{aligned}\mathbf{u}_s\cdot\nabla\times\mathbf{f} &= \lim_{\Delta V\to 0}\frac{\sum_i \mathbf{u}_s\cdot(\mathbf{n}_i\times\mathbf{f})\,\Delta S_i}{\Delta V}\\ &= \lim_{\Delta V\to 0}\frac{\sum_i \mathbf{f}\cdot(\mathbf{u}_s\times\mathbf{n}_i)\,\Delta S_i}{\Delta V}.\end{aligned} \tag{4.64}$$

The vector product $\mathbf{u}_s\times\mathbf{n}_i$ vanishes on the top and bottom surfaces of the flat box, and the contribution to the sum arises only from the side surfaces where $\mathbf{u}_s\times\mathbf{n}_i = \mathbf{u}_{\ell}$ and $\Delta S_i = \Delta\ell_i\Delta s$, while $\Delta V = \Delta s\Delta S$ ($\Delta S = \Delta A$ of Fig. 4-2). Equation (4.64), therefore, reduces to

$$\mathbf{u}_s \cdot \nabla \times \mathbf{f} = \lim_{\Delta S \to 0} \frac{\sum_i (\mathbf{f} \cdot \mathbf{u}_\ell) \Delta \ell_i}{\Delta S}$$
$$= \lim_{\Delta S \to 0} \frac{\sum \mathbf{f} \cdot \Delta \boldsymbol{\ell}_i}{\Delta S}. \tag{4.65}$$

This is the defining equation for the scalar component of $\nabla \times \mathbf{f}$ in the direction of $\mathbf{u}_s$, which can be quite arbitrary. Let us consider a contour of ΔS to be bounded by $\pm h_2 \Delta v_2 \mathbf{u}_2$ at $v_1 \pm v_1/2$, and $\pm h_1 \Delta v_1 \mathbf{u}_1$ at $v_2 \pm \Delta v_2/2$ in the general orthogonal Dupin system; then $\mathbf{u}_s = \mathbf{u}_3$ and $\Delta S = h_1 h_2 \Delta v_1 \Delta v_2$. Upon substituting these quantities into (4.65) with $\Delta \boldsymbol{\ell}_i = \pm h_2 \Delta v_2 \mathbf{u}_2$ and $\pm h_1 \Delta v_1 \mathbf{u}_1$, we obtain

$$\mathbf{u}_3 \cdot \nabla \times \mathbf{f} = \frac{1}{h_1 h_2} \left[\frac{\partial}{\partial v_1} (h_2 f_2) - \frac{\partial}{\partial v_2} (h_1 f_1) \right], \tag{4.66}$$

which is the differential expression of the scalar component of $\nabla \times \mathbf{f}$ in the direction of $\mathbf{u}_3$. By changing the orientation of $\Delta \mathbf{S}$ to $\mathbf{u}_1$ and $\mathbf{u}_2$, we can obtain the expressions for the other components. This method of finding the differential expression for $\nabla \times \mathbf{f}$ appears to be simpler than the symbolic method because we do not need the relations involving the derivatives of unit vectors. The symbolic method, however, treats the three basic functions in vector analysis on the same grounds. Furthermore, as we will reveal later, the symbolic method is very effective in systematically deriving various identities in vector analysis, as well as numerous integral theorems.

4-4 SYMBOLIC EXPRESSIONS WITH TWO FUNCTIONS AND THE PARTIAL SYMBOLIC VECTORS

Symbolic expressions with two functions are represented by $T(\nabla, f_1, f_2)$ where the two functions both can be scalar, vector, or one of each. These expressions are generated by using vector expressions consisting of three functions, of which at least one is a vector. They are listed below:

two scalars and one vector: $ab\mathbf{c}$ (4.67)

one scalar and two vectors: $a(\mathbf{b} \cdot \mathbf{c}), a(\mathbf{c} \times \mathbf{b})$ (4.68)

three vectors: $\mathbf{a} \cdot (\mathbf{b} \times \mathbf{c}), (\mathbf{a} \cdot \mathbf{c})\mathbf{b}$ or $(\mathbf{a} \cdot \mathbf{b})\mathbf{c}, \mathbf{c} \times (\mathbf{a} \times \mathbf{b})$
or $\mathbf{a} \times (\mathbf{b} \times \mathbf{c})$. (4.69)

These expressions are all well-defined functions in vector algebra, and all linear with respect to a single function. If the vector $\mathbf{c}$ in the above expressions is replaced by ∇, the symbolic vector, the following symbolic expressions are created:

$$ab\nabla = a\nabla b = b\nabla a = \nabla ab \tag{4.70}$$

$$a(\mathbf{b} \cdot \nabla) = a\nabla \cdot \mathbf{b} = \nabla \cdot a\mathbf{b} \tag{4.71}$$

$$a(\nabla \times \mathbf{b}) = \nabla \times a\mathbf{b} \tag{4.72}$$

$$\mathbf{a} \cdot (\mathbf{b} \times \nabla) = \mathbf{b} \cdot (\nabla \times \mathbf{a}) = \nabla \cdot (\mathbf{a} \times \mathbf{b}) \tag{4.73}$$

$$(\mathbf{a} \cdot \nabla)\, \mathbf{b} = (\nabla \cdot \mathbf{a})\, \mathbf{b} \tag{4.74}$$

$$(\mathbf{a} \cdot \mathbf{b})\, \nabla = \nabla\, (\mathbf{a} \cdot \mathbf{b}) \tag{4.75}$$

$$\nabla \times (\mathbf{a} \times \mathbf{b}) = (\nabla \cdot \mathbf{b})\, \mathbf{a} - (\nabla \cdot \mathbf{a})\, \mathbf{b} \tag{4.76}$$

$$\begin{aligned} \mathbf{a} \times (\mathbf{b} \times \nabla) &= (\mathbf{a} \cdot \nabla)\, \mathbf{b} - (\mathbf{a} \cdot \mathbf{b})\, \nabla \\ &= (\nabla \cdot \mathbf{a})\, \mathbf{b} - \nabla\, (\mathbf{a} \cdot \mathbf{b})\,. \end{aligned} \tag{4.77}$$

Some of the equivalent symbolic expressions are included in (4.70)–(4.77) as a result of Lemma 1. Equation (4.76) consists of two terms of the same type as given in (4.74), and (4.77) consists of one term of type (4.74), and another term of type (4.75). Thus, it is sufficient to deal with (4.70)–(4.75) as the basic expressions. The purpose of this section is to find identities involving these functions in terms of the individual functions. It is similar to the task of finding the trigonometrical identity $\sin(a + b) = \sin a \cos b + \cos a \sin b$.

The simplest method to achieve this goal is to use the differential expression of $T(\nabla, f_1, f_2)$ in Cartesian systems as stated in (4.12) without actually carrying out the differentiation. Thus, we have, for the case of two functions,

$$T(\nabla, f_1, f_2) = \sum_i \frac{\partial}{\partial x_i} T(\mathbf{a}_i, f_1, f_2) \tag{4.78}$$

where f_1 and f_2 both can be scalars, vectors, or a combination of each kind. The differentiation in (4.78) is acting on both functions. We now introduce two partial symbolic vectors, denoted by ∇_1 and ∇_2, and a lemma involving the symbolic expressions defined with respect to these two partial symbolic vectors.

A symbolic expression with a partial S vector ∇_1 is defined by

$$T(\nabla_1, f_1, f_2) = \sum_i \frac{\partial}{\partial x_i} \left[T(\mathbf{a}_i, f_1, f_2)\right]_{f_2 = c}\,. \tag{4.79}$$

The differentiation in (4.79) is acting on the function f_1 only, keeping f_2 constant. Similarly,

$$T(\nabla_2, f_1, f_2) = \sum_i \frac{\partial}{\partial x_i} \left[T(\mathbf{a}_i, f_1, f_2)\right]_{f_1 = c}\,. \tag{4.80}$$

The process is similar to the partial differentiation of a function of two independent variables, i.e.,

$$\frac{\partial f(x, y)}{\partial x} = \left[\frac{df(x, y)}{dx}\right]_{y = c}\,. \tag{4.81}$$

The name "partial S vector" was chosen because of this analogy. It is obvious that Lemma 1 is also applicable to symbolic expressions defined with a partial S vector. We now introduce the second lemma in the symbolic method.

Lemma 2. For a symbolic expression containing two functions, the following relation holds true:

$$T(\boldsymbol{\nabla}, f_1, f_2) = T(\boldsymbol{\nabla}_1, f_1, f_2) + T(\boldsymbol{\nabla}_2, f_1, f_2). \tag{4.82}$$

The proof of this lemma follows directly from the definition of these expressions, (4.79)–(4.80). This lemma can be extended to expressions with more than two functions, but in practice, it is not necessary. Let us now apply this lemma together with Lemma 1 to derive various identities in vector analysis, using (4.70)–(4.77) in that order. Since the steps involved are algebraic, most of the time we merely write the intermediate steps without comment.

1)
$$\begin{aligned} \boldsymbol{\nabla} ab &= \boldsymbol{\nabla}_a (ab) + \boldsymbol{\nabla}_b (ab) \quad \text{(Lemma 2)} \\ &= b\boldsymbol{\nabla}_a a + a\boldsymbol{\nabla}_b b, \qquad \text{(Lemma 1);} \end{aligned}$$

hence,
$$\triangledown (ab) = b \triangledown a + a \triangledown b. \tag{4.83}$$

The first term in (4.83) results from treating ab as a single function. The same steps are found in the following exercises.

2)
$$\begin{aligned} \boldsymbol{\nabla} \cdot (a\mathbf{b}) &= \boldsymbol{\nabla}_a \cdot (a\mathbf{b}) + \boldsymbol{\nabla}_b \cdot (a\mathbf{b}) \\ &= (\boldsymbol{\nabla}_a a) \cdot \mathbf{b} + a\boldsymbol{\nabla}_b \cdot \mathbf{b}; \end{aligned}$$

hence,
$$\triangledown \cdot (a\mathbf{b}) = \mathbf{b} \cdot \triangledown a + a \triangledown \cdot \mathbf{b}. \tag{4.84}$$

3)
$$\begin{aligned} \boldsymbol{\nabla} \times (a\mathbf{b}) &= \boldsymbol{\nabla}_a \times (a\mathbf{b}) + \boldsymbol{\nabla}_b \times (a\mathbf{b}) \\ &= (\boldsymbol{\nabla}_a a) \times \mathbf{b} + a\boldsymbol{\nabla}_b \times \mathbf{b}; \end{aligned}$$

hence,
$$\triangledown \times (a\mathbf{b}) = (\triangledown a) \times \mathbf{b} + a \triangledown \times \mathbf{b}. \tag{4.85}$$

4)
$$\begin{aligned} \boldsymbol{\nabla} \cdot (\mathbf{a} \times \mathbf{b}) &= \boldsymbol{\nabla}_a \cdot (\mathbf{a} \times \mathbf{b}) + \boldsymbol{\nabla}_b \cdot (\mathbf{a} \times \mathbf{b}) \\ &= \mathbf{b} \cdot (\boldsymbol{\nabla}_a \times \mathbf{a}) - \mathbf{a} \cdot (\boldsymbol{\nabla}_b \times \mathbf{b}); \end{aligned}$$

hence,
$$\triangledown \cdot (\mathbf{a} \times \mathbf{b}) = \mathbf{b} \cdot \triangledown \times \mathbf{a} - \mathbf{a} \cdot \triangledown \times \mathbf{b}. \tag{4.86}$$

5)
$$\begin{aligned} (\boldsymbol{\nabla} \cdot \mathbf{a})\,\mathbf{b} &= (\boldsymbol{\nabla}_a \cdot \mathbf{a})\,\mathbf{b} + (\boldsymbol{\nabla}_b \cdot \mathbf{a})\,\mathbf{b} \\ &= \mathbf{b}\,(\boldsymbol{\nabla}_a \cdot \mathbf{a}) + (\mathbf{a} \cdot \boldsymbol{\nabla}_b)\,\mathbf{b} \\ &= \mathbf{b} \triangledown \cdot \mathbf{a} + (\mathbf{a} \cdot \triangledown)\,\mathbf{b}. \end{aligned} \tag{4.87}$$

$(\boldsymbol{\nabla} \cdot \mathbf{a})\,\mathbf{b}$, therefore, represents the sum of two functions; it cannot be represented by a single function in a closed form.

6)
$$\boldsymbol{\nabla} (\mathbf{a} \cdot \mathbf{b}) = \mathbf{a} \times (\boldsymbol{\nabla} \times \mathbf{b}) + (\mathbf{a} \cdot \boldsymbol{\nabla})\,\mathbf{b}. \tag{4.88}$$

This equality is just another form of (4.77) because of Lemma 1. The same relation holds true when $\boldsymbol{\nabla}$ is replaced by $\boldsymbol{\nabla}_b$, which yields

$$\begin{aligned} \boldsymbol{\nabla}_b (\mathbf{a} \cdot \mathbf{b}) &= \mathbf{a} \times (\boldsymbol{\nabla}_b \times \mathbf{b}) + (\mathbf{a} \cdot \boldsymbol{\nabla}_b)\,\mathbf{b} \\ &= \mathbf{a} \times (\nabla \times \mathbf{b}) + (\mathbf{a} \cdot \nabla)\,\mathbf{b}. \end{aligned}$$

By interchanging the roles of **a** and **b**, we obtain

$$\nabla_a (\mathbf{a} \cdot \mathbf{b}) = \mathbf{b} \times (\nabla \times \mathbf{a}) + (\mathbf{b} \cdot \nabla)\, \mathbf{a}.$$

According to Lemma 2, we have

$$\nabla (\mathbf{a} \cdot \mathbf{b}) = \nabla_a (\mathbf{a} \cdot \mathbf{b}) + \nabla_b (\mathbf{a} \cdot \mathbf{b}), \tag{4.89}$$

and by treating $\mathbf{a} \cdot \mathbf{b}$ as a single function,

$$\nabla (\mathbf{a} \cdot \mathbf{b}) = \nabla (\mathbf{a} \cdot \mathbf{b}).$$

Hence,

$$\begin{aligned} \nabla (\mathbf{a} \cdot \mathbf{b}) &= \mathbf{a} \times (\nabla \times \mathbf{b}) + \mathbf{b} \times (\nabla \times \mathbf{a}) \\ &+ (\mathbf{a} \cdot \nabla)\, \mathbf{b} + (\mathbf{b} \cdot \nabla)\, \mathbf{a}. \end{aligned} \tag{4.90}$$

This exercise very clearly shows the advantage of the symbolic method in deriving such an identity.

According to (4.76),

7) $$\nabla \times (\mathbf{a} \times \mathbf{b}) = (\nabla \cdot \mathbf{b})\, \mathbf{a} - (\nabla \cdot \mathbf{a})\, \mathbf{b}. \tag{4.91}$$

The function $(\nabla \cdot \mathbf{a})\, \mathbf{b}$ is given by (4.87), and the function $(\nabla \cdot \mathbf{b})\, \mathbf{a}$ is obtained by interchanging the roles of **a** and **b** in that equation. Hence,

$$\begin{aligned} \nabla \times (\mathbf{a} \times \mathbf{b}) &= \mathbf{a}\, \nabla \cdot \mathbf{b} - \mathbf{b}\, \nabla \cdot \mathbf{a} \\ &- (\mathbf{a} \cdot \nabla)\, \mathbf{b} + (\mathbf{b} \cdot \nabla)\, \mathbf{a}. \end{aligned} \tag{4.92}$$

We have thus derived all of the identities involving two functions by an "algebraic" method without much effort, and without actually carrying out the differentiation. These identities are often assigned as exercises in existing books for students learning vector analysis, and they are quite involved if a conventional method based on differential calculus is used. For frequent reference, these identities are tabulated in Appendix B.

Finally, we will illustrate the direct application of a partial symbolic expression to derive an identity for $(\mathbf{a} \times \nabla) \times \mathbf{b}$. According to the definition of the partial symbolic expression,

$$(\mathbf{a} \times \nabla_b) \times \mathbf{b} = (\mathbf{a} \times \nabla) \times \mathbf{b}.$$

Since partial symbolic expressions also obey Lemma 1, we can write

$$(\mathbf{a} \times \nabla_b) \times \mathbf{b} = \nabla_b (\mathbf{a} \cdot \mathbf{b}) - \mathbf{a} (\nabla_b \cdot \mathbf{b}) \tag{4.93}$$

and

$$(\mathbf{b} \times \nabla_b) \times \mathbf{a} = \nabla_b (\mathbf{a} \cdot \mathbf{b}) - \mathbf{b} (\nabla_b \cdot \mathbf{a}). \tag{4.94}$$

By subtracting (4.94) from (4.93), we obtain

$$\begin{aligned} (\mathbf{a} \times \nabla_b) \times \mathbf{b} &= (\mathbf{b} \times \nabla_b) \times \mathbf{a} + \mathbf{b} (\nabla_b \cdot \mathbf{a}) - \mathbf{a} (\nabla_b \cdot \mathbf{b}) \\ &= \mathbf{a} \times (\nabla_b \times \mathbf{b}) + (\mathbf{a} \cdot \nabla_b)\, \mathbf{b} - \mathbf{a} (\nabla_b \cdot \mathbf{b}). \end{aligned}$$

Hence,

$$(\mathbf{a} \times \nabla) \times \mathbf{b} = \mathbf{a} \times (\nabla \times \mathbf{b}) + (\mathbf{a} \cdot \nabla)\,\mathbf{b} - \mathbf{a}\nabla \cdot \mathbf{b}. \tag{4.95}$$

The derivation of this formula using differential calculus is not a trivial exercise.

The function $(\mathbf{a} \cdot \nabla)\,\mathbf{b}$ occurs in several vector identities. Its differential expression in the general orthogonal coordinate system is given by

$$\begin{aligned}(\mathbf{a} \cdot \nabla)\,\mathbf{b} &= \sum_i \frac{a_i}{h_i} \frac{\partial \mathbf{b}}{\partial v_i} = \sum_i \frac{a_i}{h_i} \frac{\partial}{\partial v_i} \sum_j b_j \mathbf{u}_j \\ &= \sum_i \sum_j \frac{a_i}{h_i} \left(\frac{\partial b_j}{\partial v_i} \mathbf{u}_j + b_j \frac{\partial \mathbf{u}_j}{\partial v_i} \right).\end{aligned}$$

With the aid of (2.23) and (2.26), the derivatives of the unit vectors can be expressed in terms of the unit vectors themselves and the derivatives of the metric coefficients. The result yields

$$(\mathbf{a} \cdot \nabla)\,\mathbf{b} = \sum_i \sum_j \frac{a_i}{h_i} \frac{\partial b_j}{\partial v_i} \mathbf{u}_j + \mathbf{A} \times \mathbf{b} \tag{4.96}$$

where

$$\mathbf{A} = \frac{1}{\Omega} \sum_i \left(a_k \frac{\partial h_k}{\partial v_j} - a_j \frac{\partial h_j}{\partial v_k} \right) h_i \mathbf{u}_i \tag{4.97}$$

with $(i, j, k) = (1, 2, 3)$ in cyclic order and $\Omega = h_1 h_2 h_3$.

Before we conclude this section, it is desirable to prove Lemma 2 based on the differential form of the symbolic expression in the general orthogonal system to demonstrate once more the principle of invariance. In the general orthogonal system, (4.11), when applied to $T(\nabla, f_1, f_2)$, becomes

$$T(\nabla, f_1, f_2) = \frac{1}{\Omega} \sum_i \frac{\partial}{\partial v_i} \left[\frac{\Omega}{h_i} T(\mathbf{u}_i, f_1, f_2) \right]. \tag{4.98}$$

In (4.98), the differentiation applies not only to f_1 and f_2, but also to the factor Ω/h_i and the unit vector $\mathbf{u}_i$, which are both functions of position. Since $T(\mathbf{u}_i, f_1, f_2)$ is linear with respect to $\mathbf{u}_i$, it is proportional to $\mathbf{u}_i$, so we can combine Ω/h_i with $\mathbf{u}_i$ to form a third function $f_3 = \Omega/h_i\, \mathbf{u}_i$. For three scalar functions, we can always write

$$\begin{aligned}\frac{\partial}{\partial v_i}(f_1 f_2 f_3) &= f_2 \frac{\partial (f_1 f_3)}{\partial v_i} + f_1 \frac{\partial (f_2 f_3)}{\partial v_i} \\ &\quad - f_1 f_2 \frac{\partial f_3}{\partial v_i}.\end{aligned} \tag{4.99}$$

When the same decomposition is applied to (4.98), we can split that equation into the form

$$T(\nabla, f_1, f_2) = T(\nabla_1, f_1, f_2) + T(\nabla_2, f_1, f_2) - \frac{1}{\Omega}\sum_i \frac{\partial}{\partial v_i}\left\{\frac{\Omega}{h_i}T(\mathbf{u}_i, f_1, f_2)\right\}_{f_1, f_2 = c} \quad (4.100)$$

where

$$T(\nabla_1, f_1, f_2) = \frac{1}{\Omega}\sum_i \frac{\partial}{\partial v_i}\left[\frac{\Omega}{h_i}T(\mathbf{u}_i, f_1, f_2)\right]_{f_2 = c} \quad (4.101)$$

and

$$T(\nabla_2, f_1, f_2) = \frac{1}{\Omega}\sum_i \frac{\partial}{\partial v_i}\left[\frac{\Omega}{h_i}T(\mathbf{u}_i, f_1, f_2)\right]_{f_1 = c}. \quad (4.102)$$

The last term in (4.100) vanishes because

$$\sum_i \frac{\partial}{\partial v_i}\left(\frac{\Omega}{h_i}\mathbf{u}_i\right) = 0 \quad (4.103)$$

as a result of (2.27). Hence,

$$T(\nabla, f_1, f_2) = T(\nabla_1, f_1, f_2) + T(\nabla_2, f_1, f_2), \quad (4.104)$$

which is the statement of Lemma 2, now proved by means of (4.98), (4.101), and (4.102), all expressed in the general orthogonal system.

4-5 SYMBOLIC EXPRESSIONS WITH DOUBLE *S* VECTORS

Symbolic expressions involving two S vectors can be formed by replacing two vectors in a vector algebraic formula by two S vectors. The generating function must consist of at least two vector functions and some additional functions. The expressions so created will be denoted by $T(\nabla, \nabla, f_1, f_2, \cdots)$. A definition of these expressions invariant to the coordinate system can be given, but it is sufficient to define them specifically in the Cartesian system. The definition is given below:

$$T(\nabla, \nabla, f_1, f_2, \cdots) = \sum_i \sum_j \frac{\partial^2}{\partial x_i \partial x_j} T(\mathbf{a}_i, \mathbf{a}_j, f_1, f_2, \cdots). \quad (4.105)$$

It is understood that $(x_1, x_2, x_3) = (x, y, z)$, and $(\mathbf{a}_1, \mathbf{a}_2, \mathbf{a}_3) = (\mathbf{u}_x, \mathbf{u}_y, \mathbf{u}_z)$. This definition is obtained by applying the definition of an expression containing a single S vector in the Cartesian system, (4.12), twice. In (4.105), the functions $f_1, f_2, \cdots$ can be either scalar, vector, or a combination. Some expressions with a single function will be examined first. It is obvious that Lemma 1 also applies to (4.105).

1) $T(\nabla, \nabla, f) = (\nabla \cdot \nabla) f$; then

$$(\nabla \cdot \nabla) f = \sum_i \sum_j \frac{\partial^2}{\partial x_i \partial x_j}(\mathbf{a}_i \cdot \mathbf{a}_j) f = \sum_i \frac{\partial^2 f}{\partial x_i^2} \quad (4.106)$$

because $(\mathbf{a}_i \cdot \mathbf{a}_j) = 1$ for $i = j$ and 0 for $i \neq j$. The function found in (4.106) is called the Laplacian of f, in honor of the French mathematician Pierre Simon Laplace (1749–1827), and it is denoted by $\nabla^2 f$, i.e.,

$$\nabla \cdot \nabla f = \sum_i \frac{\partial^2 f}{\partial x_i^2} = \nabla^2 f. \tag{4.107}$$

This function can be obtained by taking the divergence of a vector function which is equal to the gradient of the scalar function f. In the Cartesian system, which we are using now,

$$\begin{aligned}\nabla \cdot (\nabla f) &= \nabla \cdot \sum_j \mathbf{a}_j \frac{\partial f}{\partial x_j} \\ &= \left(\sum_i \mathbf{a}_i \cdot \frac{\partial}{\partial x_i}\right) \sum_j \mathbf{a}_j \frac{\partial f}{\partial x_j} \\ &= \sum_i \frac{\partial^2 f}{\partial x_i^2}\end{aligned} \tag{4.108}$$

which is the same as (4.107).

2) $T(\nabla, \nabla, \mathbf{f}) = (\nabla \cdot \nabla)\mathbf{f}$; then

$$\begin{aligned}(\nabla \cdot \nabla)\mathbf{f} &= \sum_i \sum_j \frac{\partial^2}{\partial x_i \partial x_j} (\mathbf{a}_i \cdot \mathbf{a}_j)\mathbf{f} = \sum_i \frac{\partial^2 \mathbf{f}}{\partial x_i^2} \\ &= \sum_i \frac{\partial^2}{\partial x_i^2} \left(\sum_j f_j \mathbf{a}_j\right) = \sum_i \sum_j \mathbf{a}_j \frac{\partial^2 f_j}{\partial x_i^2} \\ &= \sum_j \mathbf{a}_j \nabla^2 f_j.\end{aligned} \tag{4.109}$$

The function consists of three components. The vector component in the direction of $\mathbf{a}_j$is $(\nabla^2 f_j)\mathbf{a}_j$ where $\nabla^2 f_j$ denotes the Laplacian of the scalar function f_j. Since the $\mathbf{a}_j$'s are constant vectors, it is possible to write

$$(\nabla \cdot \nabla)\mathbf{f} = \nabla^2 \mathbf{f} = \sum_j \mathbf{a}_j \nabla^2 f_j. \tag{4.110}$$

This form is true only if the function is expressed in the Cartesian system. The expression of the same function in the general orthogonal system will be discussed shortly.

3) $T(\nabla, \nabla, \mathbf{f}) = \nabla(\nabla \cdot \mathbf{f})$; then

$$\nabla(\nabla \cdot \mathbf{f}) = \sum_i \sum_j \frac{\partial^2}{\partial x_i \partial x_j} [\mathbf{a}_i(\mathbf{a}_j \cdot \mathbf{f})]$$

$$= \sum_i \sum_j \frac{\partial^2}{\partial x_i \partial x_j} (f_j \mathbf{a}_i)$$

$$= \sum_i \mathbf{a}_i \frac{\partial}{\partial x_i} \left(\sum_j \frac{\partial f_j}{\partial x_j} \right) = \nabla(\nabla \cdot \mathbf{f}). \tag{4.111}$$

4) $T(\bar{\nabla}, \bar{\nabla}, \mathbf{f}) = \bar{\nabla} \times (\bar{\nabla} \times \mathbf{f})$; then

$$\bar{\nabla} \times (\bar{\nabla} \times \mathbf{f}) = \sum_i \sum_j \frac{\partial^2}{\partial x_i \partial x_j} [\mathbf{a}_i \times (\mathbf{a}_j \times \mathbf{f})]$$

$$= \sum_i \frac{\partial}{\partial x_i} \mathbf{a}_i \times \sum_j \frac{\partial}{\partial x_j} (\mathbf{a}_j \times \mathbf{f})$$

$$= \sum_i \frac{\partial}{\partial x_i} (\mathbf{a}_i \times \nabla \times \mathbf{f}) = \nabla \times (\nabla \times \mathbf{f}). \tag{4.112}$$

From these examples, it appears again that $\bar{\nabla}$ is identical to ∇. This is, of course, not true because expressions with double S vectors obey Lemma 1, but not the corresponding expressions with double del operators. For example,

$$\bar{\nabla} \times (\bar{\nabla} \times \mathbf{f}) = -(\bar{\nabla} \times \mathbf{f}) \times \bar{\nabla} = (\mathbf{f} \times \bar{\nabla}) \times \bar{\nabla}, \tag{4.113}$$

but

$$\nabla \times (\nabla \times \mathbf{f}) \neq -(\nabla \times \mathbf{f}) \times \nabla \neq (\mathbf{f} \times \nabla) \times \nabla. \tag{4.114}$$

To demonstrate the algebraic property of $T(\bar{\nabla}, \bar{\nabla}, f)$, let us use the vector identity

$$\mathbf{a} \times (\mathbf{b} \times \mathbf{f}) = (\mathbf{a} \cdot \mathbf{f})\mathbf{b} - (\mathbf{a} \cdot \mathbf{b})\mathbf{f}$$

to generate the following symbolic expression when the vectors $\mathbf{a}$ and $\mathbf{b}$ are replaced by two symbolic vectors:

$$\bar{\nabla} \times (\bar{\nabla} \times \mathbf{f}) = (\bar{\nabla} \cdot \mathbf{f})\bar{\nabla} - (\bar{\nabla} \cdot \bar{\nabla})\mathbf{f}$$

$$= \bar{\nabla}(\bar{\nabla} \cdot \mathbf{f}) - (\bar{\nabla} \cdot \bar{\nabla})\mathbf{f}. \tag{4.115}$$

In view of (4.110)–(4.112), (4.115) is equivalent to

$$\nabla \times (\nabla \times \mathbf{f}) = \nabla(\nabla \cdot \mathbf{f}) - \nabla^2 \mathbf{f}. \tag{4.116}$$

The brackets in $\bar{\nabla} \times (\bar{\nabla} \times \mathbf{f})$ are necessary; otherwise, an expression like $\bar{\nabla} \times \bar{\nabla} \times \mathbf{f}$ is undefined. In fact, if the brackets are placed around the two S vectors, one finds

5) $$(\bar{\nabla} \times \bar{\nabla}) \times \mathbf{f} = \sum_i \sum_j \frac{\partial^2}{\partial x_i \partial x_j} (\mathbf{a}_i \times \mathbf{a}_j) \times \mathbf{f} = 0. \tag{4.117}$$

Similarly, we have

6) $$\bar{\nabla} \times \bar{\nabla} f = \nabla \times (\nabla f) = \sum_i \sum_j \frac{\partial^2}{\partial x_i \partial x_j} (\mathbf{a}_i \times \mathbf{a}_j) f = 0 \tag{4.118}$$

and

7)
$$\boldsymbol{\nabla}\cdot(\boldsymbol{\nabla}\times\mathbf{f}) = \nabla\cdot(\nabla\times\mathbf{f}) = \sum_i\sum_j \frac{\partial^2}{\partial x_i \partial x_j}\mathbf{a}_i\cdot(\mathbf{a}_j\times\mathbf{f})$$
$$= \sum_i\sum_j \frac{\partial^2}{\partial x_i \partial x_j}\mathbf{f}\cdot(\mathbf{a}_i\times\mathbf{a}_j) = 0. \tag{4.119}$$

The two identities

$$\nabla\times(\nabla f) = 0 \tag{4.120}$$

and

$$\nabla\cdot(\nabla\times\mathbf{f}) = 0 \tag{4.121}$$

correspond to (B.11) and (B.12) of Appendix B. They are often used in the formulation of engineering and physical problems.

When a symbolic expression consists of double S vectors and two functions, it is defined, in a Cartesian system, by

$$T(\boldsymbol{\nabla},\boldsymbol{\nabla},a,b) = \sum_i\sum_j \frac{\partial^2}{\partial x_i \partial x_j}T(\mathbf{a}_i,\mathbf{a}_j,a,b). \tag{4.122}$$

To simplify (4.122), we can apply Lemma 2 repeatedly to an expression with a single S vector twice, i.e.,

$$\begin{aligned} T(\boldsymbol{\nabla},\boldsymbol{\nabla},a,b) &= T(\boldsymbol{\nabla},\boldsymbol{\nabla}_a,a,b) + T(\boldsymbol{\nabla},\boldsymbol{\nabla}_b,a,b) \\ &= T(\boldsymbol{\nabla}_a,\boldsymbol{\nabla}_a,a,b) + T(\boldsymbol{\nabla}_b,\boldsymbol{\nabla}_a,a,b) \\ &\quad + T(\boldsymbol{\nabla}_a,\boldsymbol{\nabla}_b,a,b) + T(\boldsymbol{\nabla}_b,\boldsymbol{\nabla}_b,a,b) \\ &= T(\boldsymbol{\nabla}_a,\boldsymbol{\nabla}_a,a,b) + 2T(\boldsymbol{\nabla}_a,\boldsymbol{\nabla}_b,a,b) + T(\boldsymbol{\nabla}_b,\boldsymbol{\nabla}_b,a,b). \end{aligned} \tag{4.123}$$

As an example, let

$$T(\boldsymbol{\nabla},\boldsymbol{\nabla},a,b) = (\boldsymbol{\nabla}\cdot\boldsymbol{\nabla})ab = \nabla^2(ab);$$

then

$$\begin{aligned} \nabla^2(ab) &= (\boldsymbol{\nabla}_a\cdot\boldsymbol{\nabla}_a)ab + 2(\boldsymbol{\nabla}_a\cdot\boldsymbol{\nabla}_b)ab + (\boldsymbol{\nabla}_b\cdot\boldsymbol{\nabla}_b)ab \\ &= b\nabla^2 a + 2(\nabla a)\cdot(\nabla b) + a\nabla^2 b. \end{aligned} \tag{4.124}$$

We have freely used Lemma 1 in this example. In the conventional treatment, (4.124) is normally obtained by using the relations

$$\nabla(ab) = a\nabla b + b\nabla a \tag{4.125}$$

$$\nabla\cdot(a\nabla b) = a\nabla\cdot(\nabla b) + \nabla b\cdot\nabla a \tag{4.126}$$

$$\nabla\cdot(b\nabla a) = b\nabla\cdot(\nabla a) + \nabla b\cdot\nabla a \tag{4.127}$$

so that

$$\nabla \cdot (\nabla ab) = \nabla^2(ab) = b\nabla^2 a + 2(\nabla a)\cdot(\nabla b) + a\nabla^2 b. \tag{4.128}$$

Once the identity for certain functions has been obtained by using the differential form of these functions in a Cartesian system, like (4.108), (4.111), (4.112), and (4.116), we can write their differential expressions in the general orthogonal system because of the principle of invariance. Thus,

$$\begin{aligned}\nabla^2 f = \nabla\cdot(\nabla f) &= \frac{1}{\Omega}\sum_i \frac{\partial}{\partial v_i}\left(\frac{\Omega}{h_i}\mathbf{u}_i\cdot\nabla f\right)\\ &= \frac{1}{\Omega}\sum_i \frac{\partial}{\partial v_i}\left(\frac{\Omega}{h_i^2}\frac{\partial f}{\partial v_i}\right)\end{aligned} \tag{4.129}$$

$$\nabla(\nabla\cdot\mathbf{f}) = \sum_i \frac{\mathbf{u}_i}{h_i}\frac{\partial}{\partial v_i}\frac{1}{\Omega}\sum_j \frac{\partial}{\partial v_j}\left(\frac{\Omega}{h_j}f_j\right) \tag{4.130}$$

$$\nabla\times(\nabla\times\mathbf{f}) = \sum_i \frac{\mathbf{u}_i}{h_i}\times\frac{\partial}{\partial v_i}\left(\sum_j \frac{\mathbf{u}_j}{h_j}\times\frac{\partial \mathbf{f}}{\partial v_j}\right)$$

$$= \frac{1}{\Omega}\begin{vmatrix} h_1\mathbf{u}_1 & h_2\mathbf{u}_2 & h_3\mathbf{u}_3 \\ \dfrac{\partial}{\partial v_1} & \dfrac{\partial}{\partial v_2} & \dfrac{\partial}{\partial v_3} \\ \dfrac{h_1^2}{\Omega}\left[\dfrac{\partial(h_3f_3)}{\partial v_2} - \dfrac{\partial(h_2f_2)}{\partial v_3}\right] & \dfrac{h_2^2}{\Omega}\left[\dfrac{\partial(h_1f_1)}{\partial v_3} - \dfrac{\partial(h_3f_3)}{\partial v_1}\right] & \dfrac{h_3^2}{\Omega}\left[\dfrac{\partial(h_2f_2)}{\partial v_1} - \dfrac{\partial(h_1f_1)}{\partial v_2}\right]\end{vmatrix}. \tag{4.131}$$

Equation (4.131) is obtained by applying (4.41) twice. In the case of (4.116), the Laplacian of a vector function $\mathbf{f}$ in the curvilinear orthogonal system is defined by

$$\nabla^2\mathbf{f} = \nabla\cdot\nabla\mathbf{f} \tag{4.132}$$

where $\nabla\mathbf{f}$ is now a dyadic function, not a vector function. It is defined by

$$\nabla\mathbf{f} = \sum_i \frac{\mathbf{u}_i}{h_i}\frac{\partial\mathbf{f}}{\partial v_i}; \tag{4.133}$$

its meaning will be discussed in Chapter 7. The proof of (4.116) in the curvilinear orthogonal system based on (4.130)–(4.133) is found in Appendix E.

4-6 THE METHOD OF GRADIENT IN THE TRANSFORM OF UNIT VECTORS

Once the differential expressions of certain functions are available in terms of different coordinate variables, we can derive many relations from them by taking advantage of the invariance property of these functions. The method of gradient is based on this principle. We will use an example to illustrate this method.

It is known that the relationships between (x, y), two of the Cartesian variables, and (r, ϕ), two of the cylindrical variables, are

$$x = r\cos\phi \tag{4.134}$$

$$y = r\sin\phi \tag{4.135}$$

$$r = \left(x^2 + y^2\right)^{\frac{1}{2}} \tag{4.136}$$

and

$$\phi = \tan^{-1}\left(\frac{y}{x}\right). \tag{4.137}$$

By taking the gradient of (4.134) and (4.135) in the Cartesian system for x and y on the left side of these two equations, and in the cylindrical coordinate system on the right side, we obtain

$$\mathbf{u}_x = \cos\phi\mathbf{u}_r - \sin\phi\mathbf{u}_\phi \tag{4.138}$$

$$\mathbf{u}_y = \sin\phi\mathbf{u}_r + \cos\phi\mathbf{u}_\phi. \tag{4.139}$$

By doing the same for (4.136) and (4.137), but in reverse order, we obtain

$$\begin{aligned}\mathbf{u}_r &= \frac{x}{(x^2+y^2)^{\frac{1}{2}}}\mathbf{u}_x + \frac{y}{(x^2+y^2)^{\frac{1}{2}}}\mathbf{u}_y \\ &= \cos\phi\mathbf{u}_x + \sin\phi\mathbf{u}_y \end{aligned} \tag{4.140}$$

$$\begin{aligned}\frac{\mathbf{u}_\phi}{r} &= \frac{-y}{x^2+y^2}\mathbf{u}_x + \frac{x}{x^2+y^2}\mathbf{u}_y \\ \mathbf{u}_\phi &= -\sin\phi\mathbf{u}_x + \cos\phi\mathbf{u}_y. \end{aligned} \tag{4.141}$$

These relations can be derived by a geometrical method, but the method of gradient is straightforward, particularly if the orthogonal system is a more complicated one compared to the cylindrical and spherical coordinate systems. Equations (4.138)–(4.141), together with the unit vector $\mathbf{u}_z$, can be tabulated in a matrix form, as shown in Table 4-2. The table can be used in both directions. Horizontally, it gives

$$\mathbf{u}_r = \cos\phi\mathbf{u}_x - \sin\phi\mathbf{u}_y \tag{4.142}$$

$$\mathbf{u}_\phi = -\sin\phi\mathbf{u}_x + \cos\phi\mathbf{u}_y, \tag{4.143}$$

which are the same as (4.140) and (4.141). Vertically, it yields

$$\mathbf{u}_x = \cos\phi\mathbf{u}_r - \sin\phi\mathbf{u}_\phi \tag{4.144}$$

$$\mathbf{u}_y = \sin\phi\mathbf{r}_r + \cos\phi\mathbf{u}_\phi, \tag{4.145}$$

TABLE 4-2

	$\mathbf{u}_x$	$\mathbf{u}_y$	$\mathbf{u}_z$
$\mathbf{u}_r$	$\cos\phi$	$\sin\phi$	0
$\mathbf{u}_\phi$	$-\sin\phi$	$\cos\phi$	0
$\mathbf{u}_z$	0	0	1

which can be derived algebraically by solving for $\mathbf{u}_x$ and $\mathbf{u}_y$ from (4.140) and (4.141) in terms of $\mathbf{u}_r$ and $\mathbf{u}_\phi$. Each coefficient in Table 4-2 corresponds to the scalar product of the two unit vectors in the intersecting column and row; thus, $\mathbf{u}_r \cdot \mathbf{u}_x = \cos\phi$, $\mathbf{u}_\phi \cdot \mathbf{u}_x = -\sin\phi$, etc. For this reason, the same table is applicable to the transformation of the scalar components of a vector in the two systems. Since

$$\begin{aligned}\mathbf{f} &= f_x\mathbf{u}_x + f_y\mathbf{u}_y + f_z\mathbf{u}_z \\ &= f_r\mathbf{u}_r + f_\phi\mathbf{u}_\phi + f_z\mathbf{u}_z,\end{aligned}$$

then

$$\begin{aligned}f_x &= \mathbf{u}_x \cdot \mathbf{f} = (\mathbf{u}_x \cdot \mathbf{u}_r)\, f_r + (\mathbf{u}_x \cdot \mathbf{u}_\phi)\, f_\phi \\ &= \cos\phi f_r - \sin\phi f_\phi\end{aligned} \tag{4.146}$$

and

$$\begin{aligned}f_y &= \mathbf{u}_y \cdot \mathbf{f} = (\mathbf{u}_y \cdot \mathbf{u}_r)\, f_r + (\mathbf{u}_y \cdot \mathbf{u}_\phi)\, f_\phi \\ &= \sin\phi f_r + \cos\phi f_\phi.\end{aligned} \tag{4.147}$$

These relations are of the same form as (4.138) and (4.139). The transformations of the unit vectors of the orthogonal systems reviewed in Chapter 2, section 2-1 and the unit vectors in the Cartesian system are listed in Appendix A, including the cylindrical system just described.

As another example, let us consider the problem of relating $(\mathbf{u}_R, \mathbf{u}_\theta, \mathbf{u}_\phi)$ to $(\mathbf{u}_R, \mathbf{u}_\alpha, \mathbf{u}_\beta)$ of another spherical system in which the polar angle α is measured from the x axis, and the azimuthal angle β is measured with respect to the $x-y$ plane; thus,

$$x = R\sin\theta\cos\phi = R\cos\alpha \tag{4.148}$$

$$y = R\sin\theta\sin\phi = R\sin\alpha\cos\beta \tag{4.149}$$

$$z = R\cos\theta = R\sin\alpha\sin\beta. \tag{4.150}$$

We are seeking the relationships between $(\mathbf{u}_\alpha, \mathbf{u}_\beta)$ and $(\mathbf{u}_\theta, \mathbf{u}_\phi)$. The metric coefficients of the two systems are $(1, R, R\sin\theta)$ and $(1, R, R\sin\alpha)$. By taking the gradient of $\sin\theta\cos\phi = \cos\alpha$ in the two systems, we obtain

$$\frac{1}{R}\frac{\partial}{\partial\alpha}(\sin\theta\cos\phi)\,\mathbf{u}_\theta + \frac{1}{R\sin\theta}\frac{\partial}{\partial\phi}(\sin\theta\cos\phi)\,\mathbf{u}_\phi = \frac{1}{R}\frac{\partial}{\partial\alpha}(\cos\alpha)\,\mathbf{u}_\alpha. \tag{4.151}$$

Hence, $-\sin\alpha\mathbf{u}_\alpha = \cos\theta\cos\phi\mathbf{u}_\theta - \sin\phi\mathbf{u}_\phi$ or

$$\mathbf{u}_\alpha = \frac{-1}{\left(1 - \sin^2\theta\cos^2\phi\right)^{\frac{1}{2}}} \left(\cos\theta\cos\phi\mathbf{u}_\theta - \sin\phi\mathbf{u}_\phi\right). \tag{4.152}$$

By taking the gradient of

$$\cot\beta = \tan\theta\sin\phi = \left(\frac{y}{z}\right), \tag{4.153}$$

we obtain

$$\mathbf{u}_\beta = \frac{-1}{\left(1 - \sin^2\theta\cos^2\phi\right)^{\frac{1}{2}}} \left(\sin\phi\mathbf{u}_\theta + \cos\theta\cos\phi\mathbf{u}_\phi\right). \tag{4.154}$$

From (4.152) and (4.154), we can solve for $\mathbf{u}_\theta$ and $\mathbf{u}_\phi$ in terms of $\mathbf{u}_\alpha$ and $\mathbf{u}_\beta$. An alternative method is to use (4.150) and the relation

$$\tan\phi = \tan\alpha\cos\beta\left(= \frac{y}{x}\right) \tag{4.155}$$

and repeat the same operations; then we obtain

$$\mathbf{u}_\theta = \frac{-1}{\left(1 - \sin^2\alpha\sin^2\beta\right)^{\frac{1}{2}}} \left(\cos\alpha\sin\beta\mathbf{u}_\alpha + \cos\beta\mathbf{u}_\beta\right) \tag{4.156}$$

and

$$\mathbf{u}_\phi = \frac{1}{\left(1 - \sin^2\alpha\sin^2\beta\right)^{\frac{1}{2}}} \left(\cos\beta\mathbf{u}_\alpha - \cos\alpha\sin\beta\mathbf{u}_\beta\right). \tag{4.157}$$

The reader can verify these expressions by solving $\mathbf{u}_\theta$ and $\mathbf{u}_\phi$ from (4.152) and (4.154) at the expense of a tedious calculation.

These relations are very useful in antenna theory when one is interested in finding the resultant field of two linear antennas placed at the origin, with one antenna pointed in the z direction and another one pointed in the x direction. In order to calculate the resultant distant field, the individual field must be expressed in a common coordinate system, say (R, θ, ϕ) in this case. Since the field of the x-directed antenna is proportional to $\mathbf{u}_\alpha$, (4.152) can be used to combine it with the field of the z-directed antenna, whose field is proportional to $\mathbf{u}_\theta$. In fact, it is this technical problem which motivated the author to formulate the method of gradient many years ago.

Another coordinate transform dealing with the rotation of a Cartesian system should be discussed because of its practical value. Let the coordinates of a given Cartesian system be denoted by (x_1, x_2, x_3). The rotation of that system to any arbitrary orientation can be accomplished in three steps. First, we turn the (x_1, x_2) axes by an angle ϕ_1, as shown in Fig. 4-3. Let the new axes be denoted by X_1 and X_2; then

$$X_1 = x_1\cos\phi_1 + x_2\sin\phi_1 \tag{4.158}$$

$$X_2 = x_2\cos\phi_1 - x_1\sin\phi_1. \tag{4.159}$$

Now, we turn the (x_3, X_1) axes by an angle ϕ_2 where x_3 is perpendicular to both (x_1, x_2) and (X_1, X_2). The new axes in that plane will be denoted by x_3' and X_1'; then

$$x_3' = x_3 \cos\phi_2 + X_1 \sin\phi_2 \tag{4.160}$$

$$X_1' = X_1 \cos\phi_2 - x_3 \sin\phi_2. \tag{4.161}$$

Finally, we turn the (X_1', X_2) axes by an angle ϕ_3, and the new axes in that plane will be denoted by (x_1', x_2'); then

$$x_1' = X_1' \cos\phi_3 + X_2 \sin\phi_3 \tag{4.162}$$

$$x_2' = X_2 \cos\phi_3 - X_1 \sin\phi_3. \tag{4.163}$$

By solving (x_1', x_2', x_3') in terms of (x_1, x_2, x_3), we obtain

$$x_i' = \sum_j c_{ij} x_j, \qquad i = 1, 2, 3 \tag{4.164}$$

where

$$\begin{aligned}
c_{11} &= \cos\phi_1 \cos\phi_2 \cos\phi_3 - \sin\phi_1, \sin\phi_3 \\
c_{12} &= \sin\phi_1 \cos\phi_2 \cos\phi_3 + \cos\phi_1, \sin\phi_3 \\
c_{13} &= -\sin\phi_2 \cos\phi_3 \\
c_{21} &= -\cos\phi_1 \cos\phi_2 \sin\phi_3 - \sin\phi_1 \cos\phi_3 \\
c_{22} &= -\sin\phi_1 \cos\phi_2 \sin\phi_3 + \sin\phi_1 \sin\phi_3 \\
c_{23} &= \sin\phi_2 \sin\phi_3 \\
c_{31} &= \cos\phi_1 \sin\phi_2 \\
c_{32} &= \sin\phi_1 \sin\phi_2 \\
c_{33} &= \cos\phi_2.
\end{aligned}$$

By taking the gradient of (4.164) in the primed and unprimed systems, one finds that the unit vectors are governed by the same system of equations, i.e.,

$$\mathbf{a}_i' = \sum_j c_{ij} \mathbf{a}_j, \qquad i = 1, 2, 3. \tag{4.165}$$

It can be verified that

$$\mathbf{a}_i' \cdot \mathbf{a}_i' = \sum_j (c_{ij})^2 = 1, \qquad i = 1, 2, 3$$

and

$$\mathbf{a}_i' \cdot \mathbf{a}_k' = \sum_j c_{ij} c_{kj} = 0, \qquad i \neq k.$$

Since

$$c_{ik} = \mathbf{a}_i' \cdot \mathbf{a}_k,$$

the relations between the two sets of unit vectors can be tabulated in a matrix form, as in Table 4-3.

This table, like Table 4-2, can be used horizontally or vertically. When it is used horizontally, the result yields (4.165); vertically, we obtain

$$\mathbf{a}_i = \sum_j c_{ji} \mathbf{a}_j', \qquad i = 1, 2, 3.$$

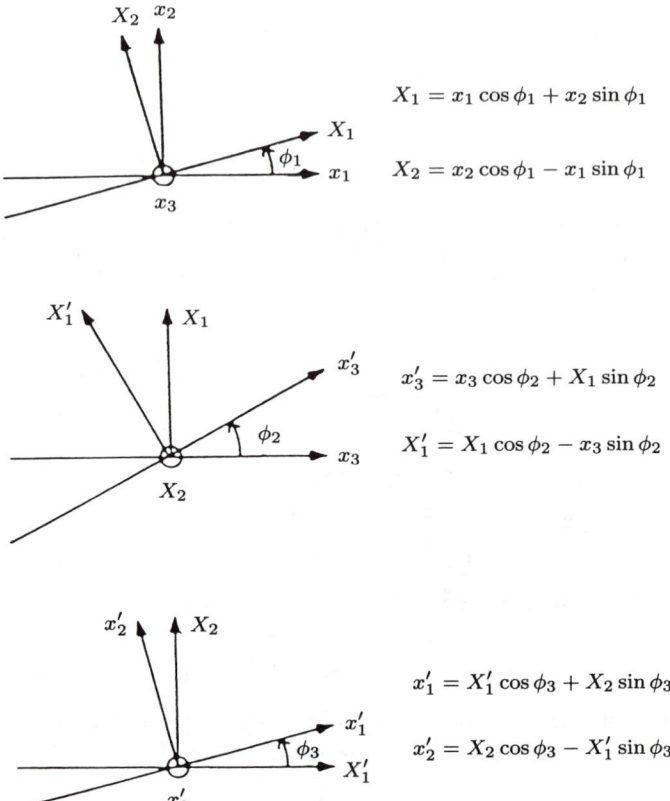

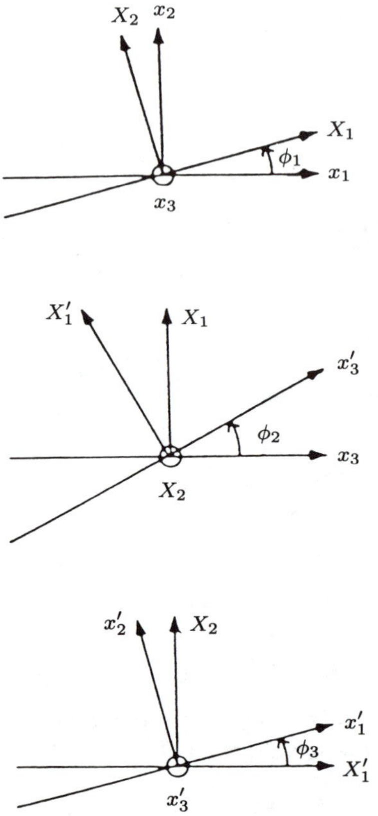

$$X_1 = x_1 \cos\phi_1 + x_2 \sin\phi_1$$

$$X_2 = x_2 \cos\phi_1 - x_1 \sin\phi_1$$

$$x_3' = x_3 \cos\phi_2 + X_1 \sin\phi_2$$

$$X_1' = X_1 \cos\phi_2 - x_3 \sin\phi_2$$

$$x_1' = X_1' \cos\phi_3 + X_2 \sin\phi_3$$

$$x_2' = X_2 \cos\phi_3 - X_1' \sin\phi_3$$

Fig. 4-3 Sequence of rotations of the axes in a Cartesian system.

TABLE 4-3

	$\mathbf{a}_1$	$\mathbf{a}_2$	$\mathbf{a}_3$
$\mathbf{a}_1'$	c_{11}	c_{12}	c_{13}
$\mathbf{a}_2'$	c_{21}	c_{22}	c_{23}
$\mathbf{a}_3'$	c_{31}	c_{32}	c_{33}

The determinant of the square matrix in Table 4-3 is numerically equal to unity because both the primed and unprimed systems are right-handed systems.

The coefficients which we have obtained are actually the directional cosines of the unit vectors, for example,

$$\begin{aligned} \mathbf{a}_1' &= \mathbf{c}_{11}\mathbf{a}_1 + c_{12}\mathbf{a}_2 + c_{13}\mathbf{a}_3 \\ &= \cos\alpha_1\mathbf{a}_1 + \cos\beta_1\mathbf{a}_2 + \cos\gamma_1\mathbf{a}_3. \end{aligned}$$

The values of the coefficients are functions of the three angles of rotation. If two of the directional cosines of one primed unit vector, say $\mathbf{a}_1'$, are specified, we still have one more degree of freedom to orient the direction of $\mathbf{a}_2'$ or $\mathbf{a}_3'$. As an example, let

$$c_{11} = \frac{1}{2},\ c_{12} = \frac{\sqrt{3}}{4};$$

then

$$c_{13} = \left(1 - c_{11}^2 - c_{12}^2\right)^{\frac{1}{2}} = \frac{3}{4}$$

so the direction of $\mathbf{a}_1'$ is specified. Suppose that we want $\mathbf{a}_3'$ to lie in the $x_1 - x_2$ plane or perpendicular to $\mathbf{a}_3$; then $c_{33} = 0$. The problem is to determine (ϕ_1, ϕ_2, ϕ_3) to meet the above specifications. The solution is embedded in the following system of equations:

$$\begin{aligned} c_{11} &= \cos\phi_1, \cos\phi_2 - \sin\phi_1 \sin\phi_3 = \frac{1}{2} \\ c_{13} &= -\sin\phi_2 \cos\phi_3 = \frac{3}{4} \\ c_{33} &= \cos\phi_2 = 0. \end{aligned}$$

One of the answers is

$$\sin\phi_1 = -\frac{2\sqrt{7}}{7},\ \phi_2 = \frac{\pi}{2},\ \cos\phi_3 = \frac{3}{4}.$$

Once these angles of rotation are known, the directions of $\mathbf{a}_2$ and $\mathbf{a}_3$ can be determined.

The method of gradient can also be used effectively to derive the expressions for the divergence operator and the curl operator in the curvilinear orthogonal system from their expressions in the Cartesian system. In the Cartesian system, the divergence operator and the curl operator are given, respectively, by (4.30) and (4.31). Using our newly created notations for these two operators, they are

$$\nabla = \sum_i \mathbf{a}_i \cdot \frac{\partial}{\partial x_i} \tag{4.166}$$

$$\nabla = \sum_i \mathbf{a}_i \times \frac{\partial}{\partial x_i}. \tag{4.167}$$

Upon applying the method of gradient to the coordinate variables x_i with $i = (1, 2, 3)$, we obtain

$$\mathbf{a}_i = \nabla x_i = \sum_j \frac{\mathbf{u}_j}{h_j} \frac{\partial x_i}{\partial v_j}, \tag{4.168}$$

and by the chain rule of differentiation,

$$\frac{\partial}{\partial x_i} = \sum_k \frac{\partial v_k}{\partial x_i} \frac{\partial}{\partial v_k}. \tag{4.169}$$

We also have the relations

$$\frac{\partial v_i}{\partial v_j} = \frac{\partial v_i}{\partial x_1}\frac{\partial x_1}{\partial v_j} + \frac{\partial v_i}{\partial x_2}\frac{\partial x_2}{\partial v_j} + \frac{\partial v_i}{\partial x_3}\frac{\partial x_3}{\partial v_j}$$

$$= \begin{cases} 1, & i = j \\ 0, & i \neq j. \end{cases} \tag{4.170}$$

Upon substituting (4.168) and (4.169) into (4.166) and (4.167), and making use of (4.170), we find

$$\nabla = \sum_i \mathbf{a}_i \cdot \frac{\partial}{\partial x_i} = \sum_i \frac{\mathbf{u}_i}{h_i} \cdot \frac{\partial}{\partial v_i} \tag{4.171}$$

$$\nabla = \sum_i \mathbf{a}_i \times \frac{\partial}{\partial x_i} = \sum_i \frac{\mathbf{u}_i}{h_i} \times \frac{\partial}{\partial v_i}. \tag{4.172}$$

These relations show that the divergence operator and the curl operator, like the del operator for the gradient, are invariant to the choice of the coordinate system.

4-7 GENERALIZED GAUSS THEOREM IN SPACE

The principal integral theorem involving a symbolic expression can be formulated based on the very definition of $T(\nabla)$, namely,

$$T(\nabla) = \lim_{\Delta V \to 0} \frac{\sum_i T(\mathbf{n}_i)\Delta S_i}{\Delta V}. \tag{4.173}$$

Equation (4.173) can be considered as a limiting form of a parent equation

$$T(\nabla) = \frac{\sum_i T(\mathbf{n}_i)\Delta S_i}{\Delta V} + \epsilon \tag{4.174}$$

where $\epsilon \to 0$ as $\Delta V \to 0$. If ΔV_j denotes a typical cell in a volume V with an enclosing surface S, then for that cell, we can write

$$T(\nabla)\Delta V_j = \sum_i T(\mathbf{n}_{ij})\Delta S_{ij} + \epsilon_j \Delta V_j \tag{4.175}$$

where ΔS_{ij} denotes an elementary area of ΔV_j, $\mathbf{n}_{ij}$ being an outward normal unit vector. By taking the Riemann sum of (4.175) with respect to j, we obtain

$$\sum_j T(\nabla)\Delta V_j = \sum_j \sum_i T(\mathbf{n}_{ij})\Delta S_{ij} + \sum_j \epsilon_j \Delta V_j. \tag{4.176}$$

Then, as $\Delta V_j \to 0$, $\epsilon_j \to 0$. By assuming $T(\nabla)$ to be continuous throughout V, we obtain

$$\iiint_V T(\nabla) dV = \oint\!\!\!\oint_S T(\mathbf{n}) dS. \tag{4.177}$$

The sign around the double integral means that the surface is closed. It is observed that the contributions of $T(\mathbf{n}_{ij})\Delta S_{ij}$ from two contacting surfaces of adjacent cells cancel each other. The only contribution results from the exterior surface where there are no neighboring cells. In (4.177), $\mathbf{n}$ denotes the outward unit normal vector to S. Normally, we should use $\mathbf{u}_n$ to denote this vector, but traditionally, $\mathbf{n}$ has been used. The same formula can be obtained by integrating (4.12) throughout V, i.e., from

$$\iiint_V T(\nabla)dV = \iiint_V \sum_i \frac{\partial T(\mathbf{a}_i)}{\partial x_i} dV. \tag{4.178}$$

The integral involving the partial derivative of $T(\mathbf{a}_i)$ with respect to x_i can be reduced to the surface integral found in (4.177). The linearity of $T(\mathbf{a}_i)$ with respect to $\mathbf{a}_i$ is a key link in that reduction. In a later section, we will give a detailed treatment of a two-dimensional version of a similar problem for a surface to demonstrate this approach. The formula that we have derived will be designated the *generalized Gauss theorem*, which converts a volume integral of $T(\nabla)$, continuous throughout V, to a surface integral evaluated at the enclosing surface S. Many of the classical theorems in vector analysis can be readily derived from this generalized theorem by a proper choice of the symbolic expression $T(\nabla)$.

1. Divergence theorem or Gauss theorem

   Let $T(\nabla) = \nabla \cdot \mathbf{f} = \nabla \cdot \mathbf{f}$. Upon substituting these quantities into (4.177), we obtain the divergence theorem or the standard Gauss theorem, named in honor of the great mathematician Carl Fredrick Gauss (1771–1855):

$$\iiint_V \nabla \cdot \mathbf{f} dV = \oiint_S (\mathbf{n} \cdot \mathbf{f}) dS. \tag{4.179}$$

2. Curl theorem

   Let $T(\nabla) = \nabla \times \mathbf{f} = \nabla \times \mathbf{f}$; then $T(\mathbf{n}) = \mathbf{n} \times \mathbf{f}$. By means of (4.177), we obtain the curl theorem:

$$\iiint_V \nabla \times \mathbf{f} dV = \oiint_S (\mathbf{n} \times \mathbf{f}) dS. \tag{4.180}$$

3. Gradient theorem

   This theorem is obtained by letting $T(\nabla) = \nabla f = \nabla f$; then $T(\mathbf{n}) = \mathbf{n} f$; hence,

$$\iiint_V \nabla f dV = \oiint_S f \mathbf{n} dS = \oiint_S f d\mathbf{S}. \tag{4.181}$$

4. Hallén's formula

If we let $T(\nabla) = (\nabla \cdot \mathbf{a})\mathbf{b}$, then $T(\mathbf{n}) = (\mathbf{n} \cdot \mathbf{a})\mathbf{b}$. Since $T(\nabla)$ consists of two functions **a** and **b**,we can apply Lemma 2 to obtain

$$\begin{aligned}(\nabla \cdot \mathbf{a})\mathbf{b} &= (\nabla_a \cdot \mathbf{a})\mathbf{b} + (\nabla_b \cdot \mathbf{a})\mathbf{b} \\ &= \mathbf{b}(\nabla_a \cdot \mathbf{a}) + (\mathbf{a} \cdot \nabla_b)\mathbf{b}.\end{aligned}$$

The rearrangement of the various terms follows Lemma 1; thus,

$$(\nabla \cdot \mathbf{a})\mathbf{b} = \mathbf{b}(\nabla \cdot \mathbf{a}) + (\mathbf{a} \cdot \nabla)\mathbf{b}. \tag{4.182}$$

By substituting (4.182) into (4.177), we obtain

$$\iiint_V [\mathbf{b}\nabla \cdot \mathbf{a} + (\mathbf{a} \cdot \nabla)\mathbf{b}]\, dV = \oiint_S (\mathbf{n} \cdot \mathbf{a})\mathbf{b}\, dS. \tag{4.183}$$

Equation (4.183), with **b** equal to $\mathbf{c}/r$, where $r = \left(x^2 + y^2 + z^2\right)^{\frac{1}{2}}$ and $\mathbf{c}$ = a constant vector which can be deleted from the resultant equation, was derived by Hallén [6], based on differential calculus carried out in a Cartesian system. We designate (4.183) as Hallén's formula for convenient identification.

The three theorems stated by (4.179)–(4.181) are closely related. In fact, it is possible to derive the divergence theorem and the curl theorem based on the gradient theorem. The derivation is given in Appendix D. The relationship between several surface theorems to be derived in Chapter 5 is also shown in that appendix.

With the vector theorems and identities at our disposal, it is of interest to give an interpretation of (2.27) based on the gradient theorem and to identify (2.28) as a vector identity.

According to (4.37), when f= constant, that equation reduces to

$$\sum_i \frac{\partial}{\partial v_i}\left(\frac{\Omega}{h_i}\mathbf{u}_i\right) = 0, \tag{4.184}$$

which is the same as (2.27), originally proved with the aid of the relationships between the derivatives of the unit vectors. From the point of view of the gradient theorem given by

$$\iiint \nabla f dV = \oiint f d\mathbf{S}$$

when f = constant, we obtain

$$\oiint d\mathbf{S} = 0, \tag{4.185}$$

which is a well-known theorem in geometry for any closed surface. Equation (4.184), therefore, can be considered as the differential form of the integral theorem for a closed surface.

In view of the definition of the curl operator given by (4.51), (2.28) is recognized as

$$\nabla \times \left(\frac{\mathbf{u}_j}{h_j}\right) = 0, \qquad j = (1, 2, 3). \tag{4.186}$$

Now,

$$\frac{\mathbf{u}_j}{h_j} = \nabla v_j;$$

hence, (4.186) is equivalent to

$$\nabla \times \nabla v_j = 0, \tag{4.187}$$

which is a valid identity according to (B.11) in Appendix B. By applying the curl theorem to the function $\mathbf{f} = \nabla v_j$, we obtain

$$\begin{aligned} \iiint \nabla \times \nabla v_j dV &= \oiint \mathbf{n} \times \nabla v_j dS \\ &= \oiint \mathbf{n} \times \mathbf{n} \frac{\partial v_j}{\partial n} dS = 0. \end{aligned} \tag{4.188}$$

Hence, (4.187) may be considered as the differential form of the integral theorem stated by (4.188).

4-8 SCALAR AND VECTOR GREEN'S THEOREMS

There are numerous theorems all bearing the name of George Green (1793–1841). We first consider Green's theorem involving scalar functions. In the Gauss theorem, stated by (4.179), if we let

$$\mathbf{f} = a\nabla b \tag{4.189}$$

where a and b are two scalar functions, then

$$\nabla \cdot \mathbf{f} = a\nabla^2 b + (\nabla a) \cdot (\nabla b), \tag{4.190}$$

which is obtained by combining identities (B.4) and (B.9) of Appendix B. Upon substituting (4.190) into (4.179) with $\mathbf{n} \cdot \mathbf{f} = a(\mathbf{n} \cdot \nabla b)$, we obtain

$$\iiint_V \left[a\nabla^2 b + (\nabla a) \cdot (\nabla b)\right] dV = \oiint_S a(\mathbf{n} \cdot \nabla b) dS. \tag{4.191}$$

Since $\mathbf{n} \cdot \nabla b$ is the scalar component of ∇b in the direction of the unit vector $\mathbf{n}$, it is equal to $\partial b/\partial n$; (4.191) is often written in the form

$$\iiint_V \left[a\nabla^2 b + (\nabla a) \cdot (\nabla b)\right] dV = \oiint_S a\frac{\partial b}{\partial n} dS. \tag{4.192}$$

For convenience, we will designate it as *the first scalar Green's theorem*.

If we let $\mathbf{f} = a\nabla b - b\nabla a$, then it is obvious that

$$\iiint_V (a\nabla^2 b - b\nabla^2 a)\, dV = \oiint_S \left(a\frac{\partial b}{\partial n} - b\frac{\partial a}{\partial n}\right) dS. \tag{4.193}$$

Equation (4.193) will be designated as *the second scalar Green's theorem*. Both (4.192) and (4.193) involve scalar functions only.

The vector Green's theorems are formulated, first, by letting

$$\mathbf{f} = \mathbf{a} \times \nabla \times \mathbf{b}. \tag{4.194}$$

In view of identity (B.6) of Appendix B, we have

$$\nabla \cdot \mathbf{f} = (\nabla \times \mathbf{b}) \cdot (\nabla \times \mathbf{a}) - \mathbf{a} \cdot (\nabla \times \nabla \times \mathbf{b}) \tag{4.195}$$

and

$$\mathbf{n} \cdot \mathbf{f} = \mathbf{n} \cdot (\mathbf{a} \times \nabla \times \mathbf{b}). \tag{4.196}$$

Upon substituting (4.195) and (4.196) into the Gauss theorem, we obtain

$$\begin{aligned} \iiint_V [(\nabla \times \mathbf{b}) \cdot (\nabla \times \mathbf{a}) - \mathbf{a} \cdot (\nabla \times \nabla \times \mathbf{b})]\, dV \\ = \oiint_S \mathbf{n} \cdot (\mathbf{a} \times \nabla \times \mathbf{b}) dS. \end{aligned} \tag{4.197}$$

Equation (4.197) is designated as *the first vector Green's theorem*. By combining (4.197) with another one with the roles of $\mathbf{a}$ and $\mathbf{b}$ interchanged or by starting with $\mathbf{f} = \mathbf{a} \times \nabla \times \mathbf{b} - \mathbf{b} \times \nabla \times \mathbf{a}$, we obtain *the second vector Green's theorem*:

$$\begin{aligned} \iiint_V [\mathbf{b} \cdot (\nabla \times \nabla \times \mathbf{a}) - \mathbf{a} \cdot (\nabla \times \nabla \times \mathbf{b})]\, dV \\ = \oiint_S \mathbf{n} \cdot (\mathbf{a} \times \nabla \times \mathbf{b} - \mathbf{b} \times \nabla \times \mathbf{a}) dS. \end{aligned} \tag{4.198}$$

The continuity of the function $\mathbf{f}$ imposed on the Gauss theorem is now carried over, for example, to the continuity of $\mathbf{a} \times \nabla \times \mathbf{b}$ in (4.198), and similarly for the other theorems.

4-9 SOLENOIDAL VECTOR, IRROTATIONAL VECTOR, AND POTENTIAL FUNCTIONS

The main purpose of this book is to treat vector analysis based on a new symbolic method. The application of vector analysis to physical problems is not covered in this treatise. However, there are several topics in introductory courses on electromagnetics and hydrodynamics involving some technical terms in vector analysis that should be introduced in a book of this nature.

When the divergence of a vector function vanishes everywhere in the entire spatial domain, such a function is called a *solenoidal vector*, and it will be denoted by $\mathbf{F}_s$ in this section. If the curl of the same vector function also vanishes everywhere, it can be proved that the function under consideration must be a constant vector. Physically, when both the divergence and the curl of a vector vanish, it means that the field has no source. In general, a solenoidal field is characterized by

$$\nabla \cdot \mathbf{F}_s = 0 \tag{4.199}$$

$$\nabla \times \mathbf{F}_s = \mathbf{f} \tag{4.200}$$

where we treat $\mathbf{f}$ as the source function responsible for producing the vector field.

When the curl of a vector vanishes but its divergence is nonvanishing, such a vector is called an *irrotational vector*, and it will be denoted by $\mathbf{F}_i$. Such a field vector is characterized by

$$\nabla \times \mathbf{F}_i = 0 \tag{4.201}$$

$$\nabla \cdot \mathbf{F}_i = f \tag{4.202}$$

where the scalar function f is treated as the source function responsible for producing the field. In electromagnetics, $\mathbf{F}_s$ corresponds to the magnetic field in magnetostatics, and $\mathbf{F}_i$ to the electric field in electrostatics. In hydrodynamics, $\mathbf{F}_i$ corresponds to the velocity field of a lemillar flow, and $\mathbf{F}_s$ to that of a vortex.

In electrodynamics, the electric and magnetic fields are coupled, and they are both functions of space and time. Their relations are governed by Maxwell's equations. For example, in air, the system of equations is

$$\nabla \times \mathbf{E} = -\mu_0 \frac{\partial \mathbf{H}}{\partial t} \tag{4.203}$$

$$\nabla \times \mathbf{H} = \mathbf{J} + \epsilon_o \frac{\partial \mathbf{E}}{\partial t} \tag{4.204}$$

$$\nabla \cdot (\epsilon_0 \mathbf{E}) = \rho \tag{4.205}$$

$$\nabla \cdot (\mu_0 \mathbf{H}) = 0 \tag{4.206}$$

$$\nabla \cdot \mathbf{J} = -\frac{\partial \rho}{\partial t} \tag{4.207}$$

where $\mathbf{J}$ and ρ denote, respectively, the current density and the charge density functions responsible for producing the electromagnetic fields $\mathbf{E}$ and $\mathbf{H}$, and μ_0 and ϵ_0 are two fundamental constants. It is seen that the magnetic field $\mathbf{H}$ is a solenoidal field, but the electric field is neither solenoidal nor irrotational, i.e., $\nabla \cdot \mathbf{E} \neq 0$ and $\nabla \times \mathbf{E} \neq 0$.

The theoretical work in electrostatics and magnetostatics is to investigate the solutions of (4.199–4.200) and (4.201–4.202) under various boundary conditions of the physical problems. In electrodynamics, the theoretical work is to study the solutions of the differential equations such as (4.203)–(4.207) for various problems. In the case of electrostatics, in view of the vector identity (B.11) of Appendix B, the electric field, now denoted by $\mathbf{E}$, can be expressed in terms of a scalar function V such that

$$\mathbf{E} = -\nabla V. \tag{4.208}$$

The negative sign in (4.208) is just a matter of tradition based on physical consideration; mathematically, it has no importance. The function V is called the *electrostatic potential function*. As a result of (4.202), with $\mathbf{F}_i$ changed to $\mathbf{E}$, we find that

$$\nabla \cdot \mathbf{E} = -\nabla \cdot \nabla V = \frac{\rho}{\epsilon_0} \tag{4.209}$$

where we have replaced the function f by ρ/ϵ_0, with ρ denoting the density function of a charge distribution and ϵ_0 a physical constant. The problem is now shifted to the study of the second-order partial differential equation

$$\nabla \cdot (\nabla V) = -\frac{\rho}{\epsilon_0} \tag{4.210}$$

which is called Poisson's equation. The operator $\nabla \cdot \nabla$ or div grad is the Laplacian operator which we have introduced in section 4-5, commonly denoted by ∇^2.

In the case of magnetostatics, in view of identity (B.12) of Appendix B, the magnetic field, now denoted by $\mathbf{H}$, replacing $\mathbf{F}_s$, can be expressed in terms of a vector function $\mathbf{A}$ such that

$$\mathbf{H} = \nabla \times \mathbf{A}. \tag{4.211}$$

$\mathbf{A}$ is called the *magnetostatic vector potential.* The function $\mathbf{f}$ in (4.200) corresponds to the density of a current distribution in magnetostatics, commonly denoted by $\mathbf{J}$. By taking the divergence of (4.200), we find that $\nabla \cdot \mathbf{f} = \nabla \cdot \mathbf{J} = 0$, which is true for a steady current. Upon substituting (4.211) into (4.200) with $\mathbf{F}_s$ and $\mathbf{f}$ replaced by $\mathbf{H}$ and $\mathbf{J}$, we obtain

$$\nabla \times \nabla \times \mathbf{A} = \mathbf{J}. \tag{4.212}$$

According to the Helmholtz theorem (Phillips [11, p. 181]), in order to determine $\mathbf{A}$, one must impose a condition on the divergence of the vector function $\mathbf{A}$ in addition to (4.211). Since

$$\nabla \times \nabla \times \mathbf{A} = -\nabla^2 \mathbf{A} + \nabla \nabla \cdot \mathbf{A}, \tag{4.213}$$

if we impose the condition

$$\nabla \cdot \mathbf{A} = 0, \tag{4.214}$$

then (4.212) becomes

$$\nabla^2 \mathbf{A} = -\mathbf{J}. \tag{4.215}$$

The condition on the divergence of $\mathbf{A}$ so imposed upon is called the gauge condition. This condition must be compatible with the resultant differential equation for $\mathbf{A}$, (4.215). By taking the divergence of that equation, we observe that $\nabla \cdot \mathbf{A}$ must be equal to zero because $\nabla \cdot \mathbf{J} = 0$. Thus, the gauge condition so imposed is indeed compatible with (4.215). The analytical work in magnetostatics now rests upon the study of the vector Poisson equation stated by (4.215) for various problems.

To solve the system of equations in electrodynamics like the ones stated by (4.203)–(4.207), we let

$$\mu_0 \mathbf{H} = \nabla \times \mathbf{A} \tag{4.216}$$

because $\mathbf{H}$ is a solenoidal vector. The function $\mathbf{A}$ is called the dynamic vector potential. Upon substituting (4.216) into (4.203), we obtain

$$\nabla \times \left(\mathbf{E} + \frac{\partial \mathbf{A}}{\partial t} \right) = 0. \tag{4.217}$$

Hence, $\mathbf{E} + \partial \mathbf{A}/\partial t$ is irrotational, so we can express it in terms of a dynamic scalar potential ϕ such that

$$\mathbf{E} + \frac{\partial \mathbf{A}}{\partial t} = -\nabla \phi. \tag{4.218}$$

Upon substituting the expressions for $\mathbf{H}$ and $\mathbf{E}$ given by (4.216) and (4.218) into (4.204), we obtain

$$\nabla \times \nabla \times \mathbf{A} = \mu_0 \mathbf{J} - \frac{1}{c^2} \left(\frac{\partial^2 \mathbf{A}}{\partial t^2} + \nabla \frac{\partial \phi}{\partial t} \right) \tag{4.219}$$

where $c = (\mu_0 \epsilon_0)^{-1/2}$ = velocity of light in free space.

In view of identity (4.213), we can impose a gauge condition on $\mathbf{A}$ such that

$$\nabla \cdot \mathbf{A} = -\frac{1}{c^2} \frac{\partial \phi}{\partial t}. \tag{4.220}$$

Then (4.219) reduces to

$$\nabla^2 \mathbf{A} + \frac{1}{c^2} \frac{\partial^2 \mathbf{A}}{\partial t^2} = -\mu_0 \mathbf{J}, \tag{4.221}$$

which is called the vector Helmholtz wave equation. By taking the divergence of (4.221) and making use of (4.207) and (4.220), we find that ϕ satisfies the following scalar Helmholtz wave equation:

$$\nabla^2 \phi + \frac{1}{c^2} \frac{\partial^2 \phi}{\partial t^2} = -\frac{\rho}{\epsilon_0}. \tag{4.222}$$

Once $\mathbf{A}$ and ϕ are known, the electromagnetic field vectors $\mathbf{E}$ and $\mathbf{H}$ can be found by using the following relations:

$$\mu_0 \mathbf{H} = \nabla \times \mathbf{A} \tag{4.223}$$

$$\mathbf{E} = -\frac{\partial \mathbf{A}}{\partial t} - \nabla \phi. \tag{4.224}$$

The method of potentials in electrodynamics is a classical method. Another approach is to deal with the equations for $\mathbf{E}$ and $\mathbf{H}$ directly. Thus, by eliminating $\mathbf{E}$ or $\mathbf{H}$ between (4.203)–(4.204), we obtain

$$\nabla \times \nabla \times \mathbf{E} + \frac{1}{c^2}\frac{\partial^2 \mathbf{E}}{\partial t^2} = -\mu_0 \frac{\partial \mathbf{J}}{\partial t} \tag{4.225}$$

$$\nabla \times \nabla \times \mathbf{H} + \frac{1}{c^2}\frac{\partial^2 \mathbf{H}}{\partial t^2} = \nabla \times \mathbf{J}. \tag{4.226}$$

These are two basic equations which can be solved by the method of dyadic Green's functions [14].

CHAPTER 5

Vector Analysis on Surface

5-1 SURFACE SYMBOLIC VECTOR AND SYMBOLIC EXPRESSION FOR A SURFACE

Vector analysis on a surface has previously been treated by Weatherburn [20]. His works are summarized by Van Bladel [1]. Most books on vector analysis do not cover this subject. The approach taken by Weatherburn is to define a two-dimensional surface operator similar to the del operator in space. Some key differential functions analogous to gradient, divergence, and curl are then introduced.

In this work, the treatment is different. We approach the analysis based on a symbolic vector method similar to the one found in Chapter 4 for vector analysis in space. A symbolic expression for a surface is defined in terms of a surface symbolic vector. Afterwards, several essential functions in vector analysis for a surface are introduced. They are different from the ones defined by Weatherburn. The relationships between the set introduced in this book and Weatherburn's set will be discussed later. Finally, it will be shown that there is an intimate relationship between the symbolic expression for a surface and the symbolic expression in space. In fact, the former can be deduced from the latter without an independent formulation. However, it is more natural to treat the vector analysis on a surface as an independent discipline first, and then point out its relationship to the vector analysis in space.

Following the symbolic method discussed in Chapter 4, we will introduce a symbolic surface vector, denoted by ∇_s, and the corresponding symbolic vector expression $T(\nabla_s)$ for a surface which is defined by

$$T(\nabla_s) = \lim_{\Delta S \to 0} \frac{\sum_i T(\mathbf{m}_i)\Delta l_i}{\Delta S} \tag{5.1}$$

where Δl_i denotes an elementary arc length of the contour enclosing ΔS, and $\mathbf{m}_i$ is the unit vector tangent to the surface and normal to its edge. The running index "i" covers the number of sides of ΔS. For a cell with four sides, "i" goes from 1 to 4. The symbolic expression is generated by replacing at least one vector in an algebraic vector expression with ∇_s. For example, $\nabla_s \times \mathbf{b}$ is created by replacing the vector "$\mathbf{a}$" in $\mathbf{a} \times \mathbf{b}$ with ∇_s. The expression defined by (5.1) is invariant to, or independent of the choice of, the coordinates on the surface in the general Dupin system. It is recalled that the choice of (v_1, v_2) and the corresponding tangential unit vectors $(\mathbf{u}_1, \mathbf{u}_2)$ is quite arbitrary. To find the differential expression based on (5.1) in the general orthogonal Dupin system, let the sides of the surface cell be located at $v_1 \pm (\Delta v_1/2)$ and $v_2 \pm (\Delta v_2/2)$, with the corresponding unit normal vectors $\pm\mathbf{u}_1$ and $\pm\mathbf{u}_2$ located at these positions. The value of $\mathbf{u}_1$ evaluated at $v_1 + (\Delta v_1/2)$ is not equal to the value of the same unit vector evaluated at $v_1 - (\Delta v_1/2)$. The same is true for the metric coefficients h_1 and h_2. The area of the elementary surface ΔS is equal to $h_1 h_2 \Delta v_1 \Delta v_2$. Figure 5-1 shows the configuration of the cell. By substituting these quantities into (5.1) and taking the limit, one finds

$$T(\nabla_s) = \frac{1}{h_1 h_2}\left\{\frac{\partial}{\partial v_1}\left[h_2 T(\mathbf{u}_1)\right] + \frac{\partial}{\partial v_2}\left[h_1 T(\mathbf{u}_2)\right]\right\}. \tag{5.2}$$

For a plane surface located in the x-y plane in a Cartesian system,

$$T(\nabla_s) = \frac{\partial}{\partial x} T(\mathbf{u}_x) + \frac{\partial}{\partial y} T(\mathbf{u}_y). \tag{5.3}$$

This is the only case where $T(\nabla_s)$ can be expressed conveniently in a Cartesian system. In general, Cartesian variables are not the proper ones to describe the function $T(\nabla_s)$ for a curved surface. From the definition of $T(\nabla_s)$ given by (5.1) and its differential form stated by (5.2), it is obvious that Lemma 1, introduced at the end of Chapter 4, section 4-1, is also applicable to $T(\nabla_s)$ because $T(\mathbf{m}_i)$, $T(\mathbf{u}_1)$, and $T(\mathbf{u}_2)$ all have the proper form of a vector expression. By means of (5.2), it is now possible to derive the differential expression of some key functions in the vector analysis for a surface, analogous to the gradient, the divergence, and the curl in space.

5-2 SURFACE GRADIENT, SURFACE DIVERGENCE, AND SURFACE CURL

For convenience, we repeat here the differential expression for $T(\nabla_s)$ expressed in the general Dupin system:

$$T(\nabla_s) = \frac{1}{h_1 h_2}\left\{\frac{\partial}{\partial v_1}\left[h_2 T(\mathbf{u}_1)\right] + \frac{\partial}{\partial v_2}\left[h_1 T(\mathbf{u}_2)\right]\right\}. \tag{5.4}$$

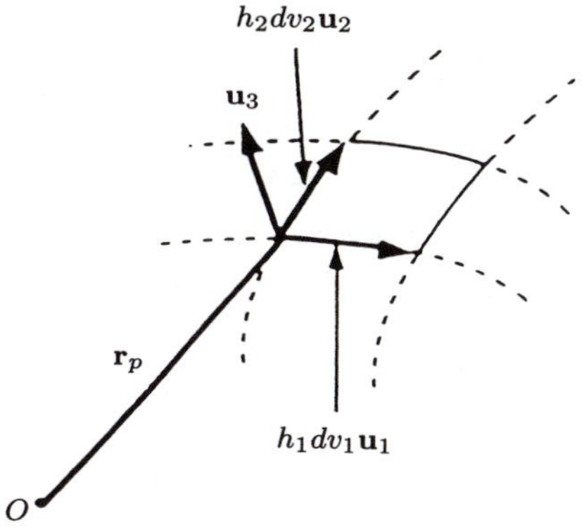

Fig. 5-1 Cell on a surface in the general Dupin coordinate system.

1. Surface Gradient

If we let $T(\nabla_s) = \nabla_s f$ where f is a scalar function of position, then $T(\mathbf{u}_1) = f\mathbf{u}_1$, $T(\mathbf{u}_2) = f\mathbf{u}_2$. Upon substituting these quantities into (5.4), we obtain

$$T(\nabla_s) = \frac{1}{h_1 h_2}\left[\frac{\partial}{\partial v_1}(h_2 f\mathbf{u}_1) + \frac{\partial}{\partial v_2}(h_1 f\mathbf{u}_2)\right]$$

$$= \frac{1}{h_1 h_2}\left[h_2 f\frac{\partial \mathbf{u}_1}{\partial v_1} + \left(f\frac{\partial h_2}{\partial v_1} + h_2\frac{\partial f}{\partial v_1}\right)\mathbf{u}_1 + h_1 f\frac{\partial \mathbf{u}_2}{\partial v_2} + \left(f\frac{\partial h_1}{\partial v_2} + h_1\frac{\partial f}{\partial v_2}\right)\mathbf{u}_2\right].$$

By making use of (2.26), with $h_3 = 1$, we can express the derivatives of $\mathbf{u}_1$ and $\mathbf{u}_2$ in terms of $(\mathbf{u}_1, \mathbf{u}_2, \mathbf{u}_3)$, which yields

$$T(\nabla_s) = \frac{1}{h_1 h_2}\left[-h_2 f\left(\frac{1}{h_2}\frac{\partial h_1}{\partial v_2}\mathbf{u}_2 + \frac{\partial h_1}{\partial v_3}\mathbf{u}_3\right) + \left(f\frac{\partial h_2}{\partial v_1} + h_2\frac{\partial f}{\partial v_1}\right)\mathbf{u}_1 - h_1 f\left(\frac{1}{h_1}\frac{\partial h_2}{\partial v_1}\mathbf{u}_1 + \frac{\partial h_2}{\partial v_3}\mathbf{u}_3\right) + \left(f\frac{\partial h_1}{\partial v_2} + h_1\frac{\partial f}{\partial v_2}\right)\mathbf{u}_2\right].$$

Some of the terms cancel each other. The net result is

$$\nabla_s f = \frac{1}{h_1}\frac{\partial f}{\partial v_1}\mathbf{u}_1 + \frac{1}{h_2}\frac{\partial f}{\partial v_2}\mathbf{u}_2 - \left(\frac{1}{h_1}\frac{\partial h_1}{\partial v_3} + \frac{1}{h_2}\frac{\partial h_2}{\partial v_3}\right) f\mathbf{u}_3. \qquad (5.5)$$

The coefficient in front of $f\mathbf{u}_3$ can be written in several different forms. If we denote the product $h_1 h_2$ by H, which is equal to $\Omega(= h_1 h_2 h_3)$ with $h_3 = 1$, then

$$\frac{1}{h_1}\frac{\partial h_1}{\partial v_3} + \frac{1}{h_2}\frac{\partial h_2}{\partial v_3} = \frac{1}{H}\frac{\partial H}{\partial v_3} = -\left(\frac{1}{R_1} + \frac{1}{R_2}\right) = -J \tag{5.6}$$

where R_1 and R_2 denote the principal radii of curvature of the surface as stated by (2.37) and (2.40). In the language of differential geometry, the parameter J is called the first curvature. Actually, it is the sum of two Gaussian curvatures. The differential function which we have just derived is designated as the *surface gradient* of f, and it will be denoted by$\nabla_s f$, which can now be written in the form

$$\nabla_s f = \frac{1}{h_1}\frac{\partial f}{\partial v_1}\mathbf{u}_1 + \frac{1}{h_2}\frac{\partial f}{\partial v_2}\mathbf{u}_2 + Jf\mathbf{u}_3. \tag{5.7}$$

It is observed that the first two terms of (5.7) are identical to the first two terms of the three-dimensional gradient ∇f when the latter is expressed in the general Dupin coordinate system with metric coefficients $(h_1, h_2, 1)$, i.e.,

$$\nabla f = \frac{1}{h_1}\frac{\partial f}{\partial v_1}\mathbf{u}_1 + \frac{1}{h_2}\frac{\partial f}{\partial v_2}\mathbf{u}_2 + \frac{\partial f}{\partial v_3}\mathbf{u}_3. \tag{5.8}$$

The connection between $T(\nabla)$ and $T(\nabla_s)$, in general, will be discussed in great detail later.

2. Surface Divergence

If we let $T(\nabla_s) = \nabla_s \cdot \mathbf{f}$, then $T(\mathbf{u}_1) = \mathbf{u}_1 \cdot \mathbf{f} = f_1, T(\mathbf{u}_2) = \mathbf{u}_2 \cdot \mathbf{f} = f_2$; hence,

$$\nabla_s \cdot \mathbf{f} = \frac{1}{h_1 h_2}\left[\frac{\partial}{\partial v_1}(h_2 f_1) + \frac{\partial}{\partial v_2}(h_1 f_2)\right], \tag{5.9}$$

which can be written in the form

$$\nabla_s \cdot \mathbf{f} = \frac{1}{H}\sum_{i=1}^{2}\frac{\partial}{\partial v_i}\left(\frac{H}{h_i}f_i\right). \tag{5.10}$$

The function so obtained is designated as the *surface divergence* of $\mathbf{f}$, and it will be denoted by $\nabla_s \cdot \mathbf{f}$; thus,

$$\nabla_s \cdot \mathbf{f} = \frac{1}{H}\sum_{i=1}^{2}\frac{\partial}{\partial v_i}\left(\frac{H}{h_i}f_i\right). \tag{5.11}$$

It is observed that this function corresponds to the first two terms of the three-dimensional divergence when the latter is expressed in the general Dupin coordinate system, $h_3 = 1$, i.e.,

$$\nabla \cdot \mathbf{f} = \frac{1}{H}\sum_{i=1}^{3}\frac{\partial}{\partial v_i}\left(\frac{H}{h_i}f_i\right). \tag{5.12}$$

We must now be very cautious about the index number. In (5.11), "i" runs from 1 to 2, and in (5.12), it runs from 1 to 3.

3. Surface Curl

Finally, if we let $T(\nabla_s) = \nabla_s \times \mathbf{f}$, then $T(\mathbf{u}_1) = \mathbf{u}_1 \times \mathbf{f} = f_2\mathbf{u}_3 - f_3\mathbf{u}_2$, and $T(\mathbf{u}_2) = \mathbf{u}_2 \times \mathbf{f} = f_3\mathbf{u}_1 - f_1\mathbf{u}_3$. Upon substituting them into (5.2), we obtain

$$\nabla_s \times \mathbf{f} = \frac{1}{h_1 h_2}\left\{\frac{\partial}{\partial v_1}[h_2(f_2\mathbf{u}_3 - f_3\mathbf{u}_2)] + \frac{\partial}{\partial v_2}[h_1(f_3\mathbf{u}_1 - f_1\mathbf{u}_3)]\right\}. \tag{5.13}$$

Again, the derivatives of the unit vectors in (5.13) can be expressed in terms of the unit vectors by making use of (2.23). After a tedious reduction and a regrouping of the terms, we obtain

$$\nabla_s \times \mathbf{f} = \frac{1}{h_1 h_2}\left\{\left[h_1\frac{\partial f_3}{\partial v_2} + h_2 f_2 \frac{\partial h_1}{\partial v_3}\right]\mathbf{u}_1 - \left[h_2\frac{\partial f_3}{\partial v_1} + h_1 f_1 \frac{\partial h_2}{\partial v_3}\right]\mathbf{u}_2 + \left[\frac{\partial(h_2 f_2)}{\partial v_1} - \frac{\partial(h_1 f_1)}{\partial v_2}\right]\mathbf{u}_3\right\}. \tag{5.14}$$

The function so obtained is designated as the *surface curl*, and it will be denoted by $\nabla_s \times \mathbf{f}$. Now, the curl function in three-dimensional space when expressed in the general Dupin orthogonal system is

$$\nabla \times \mathbf{f} = \frac{1}{h_1 h_2}\left\{\left[h_1\frac{\partial f_3}{\partial v_2} - h_1\frac{\partial(h_2 f_2)}{\partial v_3}\right]\mathbf{u}_1 + \left[h_2\frac{\partial(h_1 f_1)}{\partial v_3} - h_2\frac{\partial f_3}{\partial v_1}\right]\mathbf{u}_2 + \left[\frac{\partial(h_2 f_2)}{\partial v_1} - \frac{\partial(h_1 f_1)}{\partial v_2}\right]\mathbf{u}_3\right\}. \tag{5.15}$$

It is seen that the third term in (5.14) and the third term in (5.15) are identical, but the relationship between the other terms is not obvious. We will defer the comparison between $\nabla_s \times \mathbf{f}$ and $\nabla \times \mathbf{f}$ until the general relation between $T(\nabla_s)$ and $T(\nabla)$ is discussed. In summary, we have defined three basic functions in the vector analysis on a surface based on a symbolic method: they are $\nabla_s f$, $\nabla_s \cdot \mathbf{f}$, and $\nabla_s \times \mathbf{f}$. For the time being, we do not attach significance to the symbol ∇_s; it is used merely to denote the surface gradient, the surface divergence, and the surface curl. In particular, we should not treat ∇_s in the same way as ∇_s, the latter being a surface symbolic vector. We have pointed out before that Lemma 1 applies to $T(\nabla_s)$, for example, $\nabla_s \cdot \mathbf{f} = \mathbf{f} \cdot \nabla_s$; but $f\nabla_s$ has no meaning whatsoever at this moment. Later, the symbol ∇_s will be revealed as a differential–algebraic operator.

5-3 RELATIONSHIP BETWEEN THE VOLUME AND SURFACE SYMBOLIC EXPRESSIONS

Although the surface symbolic vector ∇_s and the symbolic expression $T(\nabla_s)$ involving ∇_s as defined by (5.1) and its differential form by (5.2) appear to be independent of ∇ and $T(\nabla)$, actually they are intimately related. If we express $T(\nabla)$ in the general Dupin system ($h_3 = 1$), then (4.11) becomes

$$T(\nabla) = \frac{1}{H}\sum_{i=1}^{3} \frac{\partial}{\partial v_i}\left[\frac{H}{h_i}T(\mathbf{u}_i)\right] \tag{5.16}$$

where $H = h_1 h_2$. The differential expression of $T(\nabla_s)$ as given by (5.2) can be written in the form

$$T(\nabla_s) = \frac{1}{H}\sum_{i=1}^{2} \frac{\partial}{\partial v_i}\left[\frac{H}{h_i}T(\mathbf{u}_i)\right]. \tag{5.17}$$

It is obvious that the first two terms of (5.16) are exactly the same as $T(\nabla_s)$; hence,

$$T(\nabla) = T(\nabla_s) + \frac{1}{H}\frac{\partial}{\partial v_3}\left[HT(\mathbf{u}_3)\right]. \tag{5.18}$$

Equation (5.18), therefore, can be used to find $T(\nabla_s)$ once $T(\nabla)$ is known or $T(\nabla_s)$ can be defined as the sum of the first two terms of $T(\nabla)$. From this point of view, $T(\nabla_s)$ is not an independent function, and ∇_s is not an independent symbolic vector.

The last term of (5.18) can be written in the form

$$\begin{aligned}\frac{1}{H}\frac{\partial}{\partial v_3}\left[HT(\mathbf{u}_3)\right] &= \frac{\partial T(\mathbf{u}_3)}{\partial v_3} + \frac{1}{H}\frac{\partial H}{\partial v_3}T(\mathbf{u}_3)\\ &= \frac{\partial T(\mathbf{u}_3)}{\partial v_3} - JT(\mathbf{u}_3).\end{aligned} \tag{5.19}$$

Equation (5.18) is, therefore, equivalent to

$$T(\nabla) = T(\nabla_s) + \frac{\partial T(\mathbf{u}_3)}{\partial v_3} - JT(\mathbf{u}_3). \tag{5.20}$$

By using the expression of ∇f, $\nabla \cdot \mathbf{f}$, and $\nabla \times \mathbf{f}$ given by (5.8), (5.12), (5.15) and the expressions of $\nabla_s f$, $\nabla_s \cdot \mathbf{f}$, and $\nabla_s \times \mathbf{f}$ given by (5.7), (5.11), (5.14), it can be easily verified that (5.20) is indeed satisfied when we let $T(\nabla)$ equal ∇f, $\nabla \cdot \mathbf{f}$, $\nabla \times \mathbf{f}$, respectively.

5-4 RELATIONSHIP BETWEEN WEATHERBURN'S SURFACE FUNCTIONS AND THE FUNCTIONS DEFINED BY THE SYMBOLIC METHOD

In the classical work of Weatherburn [20], the surface gradient, the surface divergence, and the surface curl are defined differently. His work will be reviewed and compared to the surface functions derived by the symbolic method. Following Weatherburn, a two-dimensional differential operator, denoted by ∇_t, is defined by

$$\nabla_t = \sum_{i=1}^{2} \frac{\mathbf{u}_i}{h_i} \frac{\partial}{\partial v_i}, \tag{5.21}$$

which will be designated as the *transversal del operator*. It represents the transversal part of the three-dimensional del operator

$$\nabla = \sum_{i=1}^{3} \frac{\mathbf{u}_i}{h_i} \frac{\partial}{\partial v_i}. \tag{5.22}$$

In the general Dupin system, $h_3 = 1$, the del operator becomes

$$\nabla = \nabla_t + \mathbf{u}_3 \frac{\partial}{\partial v_3}. \tag{5.23}$$

When we apply the transversal del operator to a scalar function f, we obtain

$$\begin{aligned} \nabla_t f &= \left(\sum_{i=1}^{2} \frac{\mathbf{u}_i}{h_i} \frac{\partial}{\partial v_i} \right) f \\ &= \frac{1}{h_1} \frac{\partial f}{\partial v_1} \mathbf{u}_1 + \frac{1}{h_2} \frac{\partial f}{\partial v_2} \mathbf{u}_2. \end{aligned} \tag{5.24}$$

Of course, we assume that the operation is distributive, i.e.,

$$\left(\sum_{i=1}^{2} \frac{\mathbf{u}_i}{h_i} \frac{\partial}{\partial v_i} \right) f = \sum_{i=1}^{2} \frac{\mathbf{u}_i}{h_i} \frac{\partial f}{\partial v_i}.$$

The function described by (5.24) is designated as the *surface gradient* of f by Weatherburn. The notation $\text{grad}_s f$ is used by Van Bladel [1]. Incidentally, in Weatherburn's original work, he simply uses the del symbol "∇," which is very confusing indeed. We have now adopted a new notation ∇_t so that Weatherburn's surface gradient is written in the manner of (5.24).

Weatherburn defines the surface divergence, denoted by $\nabla_t \cdot \mathbf{f}$, as

$$\nabla_t \cdot \mathbf{f} = \sum_{i=1}^{2} \frac{\mathbf{u}_i}{h_i} \cdot \left(\frac{\partial}{\partial v_i} \sum_{j=1}^{3} f_j \mathbf{u}_j \right). \tag{5.25}$$

Then

$$\begin{aligned} \nabla_t \cdot \mathbf{f} &= \frac{\mathbf{u}_1}{h_1} \cdot \frac{\partial}{\partial v_1} (f_1 \mathbf{u}_1 + f_2 \mathbf{u}_2 + f_3 \mathbf{u}_3) \\ &\quad + \frac{\mathbf{u}_2}{h_2} \cdot \frac{\partial}{\partial v_2} (f_1 \mathbf{u}_1 + f_2 \mathbf{u}_2 + f_3 \mathbf{u}_3). \end{aligned} \tag{5.26}$$

By making use of (2.23) and (2.26), with $h_3 = 1$ in these formulas, we can express the derivatives of the unit vectors in terms of the unit vectors themselves. After simplifying the result, we obtain

$$\nabla_t \cdot \mathbf{f} = \frac{1}{h_1 h_2}\left[\frac{\partial(h_2 f_1)}{\partial v_1} + \frac{\partial(h_1 f_2)}{\partial v_2}\right] - J f_3 \tag{5.27}$$

where J is the coefficient for the first curvature of the surface as defined by (5.6). The function represented by (5.27) is the *surface divergence* of $\mathbf{f}$ defined by Weatherburn. Van Bladel uses the notation $\text{div}_s\mathbf{f}$ for this function.

Finally, Weatherburn defines the surface curl, denoted by $\nabla_t \times \mathbf{f}$, as

$$\nabla_t \times \mathbf{f} = \sum_{i=1}^{2} \frac{\mathbf{u}_i}{h_i} \times \left(\frac{\partial}{\partial v_i} \sum_{j=1}^{3} f_j \mathbf{u}_j\right). \tag{5.28}$$

We obtain, after a lengthy calculation and a rearrangement of terms,

$$\begin{aligned}\nabla_t \times \mathbf{f} = \frac{1}{h_1 h_2}\Bigg\{&\left[h_1 \frac{\partial f_3}{\partial v_2} - f_2 h_1 \frac{\partial h_2}{\partial v_3}\right]\mathbf{u}_1 \\ &+ \left[-h_2 \frac{\partial f_3}{\partial v_1} + f_1 h_2 \frac{\partial h_1}{\partial v_3}\right]\mathbf{u}_2 \\ &+ \left[\frac{\partial(h_2 f_2)}{\partial v_1} - \frac{\partial(h_1 f_1)}{\partial v_2}\right]\mathbf{u}_3\Bigg\}.\end{aligned} \tag{5.29}$$

Weatherburn used this function to define his surface curl of $\mathbf{f}$. The notation used by Van Bladel for this function is $\text{curl}_s\mathbf{f}$.

By comparing Weatherburn's functions to the surface functions derived by the symbolic method and our definition for these functions, we find some very simple relationships between the two sets. They are

$$\nabla_s f = \nabla_t f + J(\mathbf{u}_3 f) \tag{5.30}$$

$$\nabla_s \cdot \mathbf{f} = \nabla_t \cdot \mathbf{f} + J(\mathbf{u}_3 \cdot \mathbf{f}) \tag{5.31}$$

$$\nabla_s \times \mathbf{f} = \nabla_t \times \mathbf{f} + J(\mathbf{u}_3 \times \mathbf{f}). \tag{5.32}$$

Equation (5.30) clearly shows that "∇_s" is not only a symbol used for our surface functions, but it is indeed a combined differential and algebraic operator defined by

$$\nabla_s = \nabla_t + J\mathbf{u}_3 = \sum_{i=1}^{2} \frac{\mathbf{u}_i}{h_i}\frac{\partial}{\partial v_i} + J\mathbf{u}_3. \tag{5.33}$$

Actually, this formula can be derived at the very beginning of section 5-1 after the differential expression of $T(\nabla_s)$ has been obtained in the general Dupin system as given by (5.2). To show this derivation, let us introduce an abstract format for $T(\nabla_s)$ in the form

$$T(\nabla_s) = \nabla_s * \tilde{g}, \tag{5.34}$$

which is similar to (4.43), except that it is defined with respect to ∇_s instead of ∇. Then, we have

$$\begin{aligned}
\nabla_s * \tilde{g} &= \frac{1}{h_1 h_2} \sum_{i=1}^{2} \frac{\partial}{\partial v_i} \left(\frac{H}{h_i} \mathbf{u}_i * \tilde{g} \right) \\
&= \frac{1}{h_1 h_2} \sum_{i=1}^{2} \frac{\partial}{\partial v_i} \left(\frac{H}{h_i} \mathbf{u}_i \right) * \tilde{g} + \frac{1}{h_1 h_2} \sum_{i=1}^{2} \frac{H}{h_i} \mathbf{u}_i * \frac{\partial \tilde{g}}{\partial v_i} \\
&= \frac{1}{h_1 h_2} \left[h_2 \frac{\partial \mathbf{u}_1}{\partial v_1} + \mathbf{u}_1 \frac{\partial h_2}{\partial v_1} + h_1 \frac{\partial \mathbf{u}_2}{\partial v_2} + \mathbf{u}_2 \frac{\partial h_1}{\partial v_2} \right] * \tilde{g} + \sum_{i=1}^{2} \frac{\mathbf{u}_i}{h_i} * \frac{\partial \tilde{g}}{\partial v_i} \\
&= \left(\frac{1}{h_1} \frac{\partial h_1}{\partial v_3} + \frac{1}{h_2} \frac{\partial h_2}{\partial v_3} \right) \mathbf{u}_3 * \tilde{g} + \sum_{i=1}^{2} \frac{\mathbf{u}_i}{h_i} * \frac{\partial \tilde{g}}{\partial v_i} \\
&= (\nabla_t + J\mathbf{u}_3) * \tilde{g} = \nabla_s * \tilde{g}
\end{aligned} \tag{5.35}$$

where we have used (2.26) to simplify the derivatives of the unit vectors. The origin of the differential–algebraic operator ∇_s is now quite clear. It is not merely a symbol; it is a well-defined operator.

As far as the surface functions are concerned, once the relationships between the two sets are known, it is a matter of personal preference as to which set should be considered as the standard surface functions. In a subsequent section dealing with integral theorems, it will be evident that the set derived from the present method, i.e., $\nabla_s f$, $\nabla_s \cdot \mathbf{f}$, and $\nabla_s \times \mathbf{f}$,or in general, $T(\nabla_s)$, is much more convenient to formulate the generalized Gauss theorem for a surface. We may also inject a remark that in electromagnetic theory, the equation of continuity (the law of conservation of charge) relating the surface current density $\mathbf{J}_s$ and the time rate of the change of the surface charge density ρ_s is described by

$$\nabla_s \cdot \mathbf{J}_s = -\frac{\partial \rho_s}{\partial t} \tag{5.36}$$

when the surrounding medium has no loss. Here it is $\nabla_s \cdot \mathbf{J}$, not $\nabla_t \cdot \mathbf{J}$ that enters the formulation. On the other hand, for the rate of change of a scalar function on a surface in a direction tangent to the surface, both $\nabla_s f$ and $\nabla_t f$ produce the same result:

$$\frac{\partial f}{\partial s} = \mathbf{u}_s \cdot \nabla_s f = \mathbf{u}_s \cdot \nabla_t f. \tag{5.37}$$

The vector component $(\partial f / \partial v_3)\mathbf{u}_3$ in $\nabla_s f$ does not affect the value of $\partial f / \partial s$. For the curl function, one finds

$$\mathbf{u}_3 \cdot \nabla_s \times \mathbf{f} = \mathbf{u}_3 \cdot \nabla_t \times \mathbf{f} = \mathbf{u}_3 \cdot \nabla \times \mathbf{f}, \tag{5.38}$$

an identity to be used later.

5-5 GENERALIZED GAUSS THEOREM FOR A SURFACE

By integrating the differential expression for $T(\nabla_s)$, (5.2), on an open surface S with contour L, we have

$$\iint_S T(\nabla_s)dS = \iint_S \sum_{i=1}^{2} \frac{\partial}{\partial v_i}\left[\frac{H}{h_i}T(\mathbf{u}_i)\right] dv_1 dv_2 \tag{5.39}$$

where $dS = h_1 h_2 dv_1 dv_2 = H dv_1 dv_2$. We assume that $T(\mathbf{u}_i)$ is continuous throughout S. The integrals in (5.39) can be carried out as follows:

$$\begin{aligned}\iint_S \frac{\partial}{\partial v_1}[h_2 T(\mathbf{u}_1)] dv_1 dv_2 &= \int_{v_{2\,\min}}^{v_{2\,\max}} [h_2 T(\mathbf{u}_1)]_{P_1}^{P_2} dv_2 \\ &= \int_{L_2} h_2 T(\mathbf{u}_1) dv_2 - \int_{-L_1} h_2 T(\mathbf{u}_1) dv_2 \\ &= \oint_L h_2 T(\mathbf{u}_1) dv_2 \end{aligned}\tag{5.40}$$

where the locations (P_1, P_2), the segments L_1, L_2, and the two extreme values $v_{2\,\min}$, $v_{2\,\max}$ are shown in Fig. 5-2(a). Similarly,

$$\begin{aligned}\iint_S \frac{\partial}{\partial v_2}[h_1 T(\mathbf{u}_2)] dv_1 dv_2 &= \int_{v_{1\,\min}}^{v_{1\,\max}} [h_1 T(\mathbf{u}_2)]_{P_3}^{P_4} dv_1 \\ &= \int_{-L_4} h_1 T(\mathbf{u}_2) dv_1 - \int_{L_3} h_1 T(\mathbf{u}_2) dv_1 \\ &= -\oint_L h_1 T(\mathbf{u}_2) dv_1 \end{aligned}\tag{5.41}$$

where the locations (P_3, P_4), the segments L_3, L_4, and the two extreme values $v_{1\,\min}$, $v_{1\,\max}$ are shown in Fig. 5-2(b). Hence,

$$\iint_S T(\nabla_s) dS = \oint_L [h_2 T(\mathbf{u}_1) dv_2 - h_1 T(\mathbf{u}_2) dv_1]. \tag{5.42}$$

Because $T(\mathbf{u}_i)$ is linear with respect to $\mathbf{u}_i$ with $i = 1, 2$, the integrand in the line integral is proportional to

$$\mathbf{u}_1 h_2 dv_2 - \mathbf{u}_2 h_1 dv_1, \tag{5.43}$$

which can be simplified. Let us consider a segment of the contour L_1, which is the edge of S. In the tangential plane containing $\mathbf{u}_1$ and $\mathbf{u}_2$ at a typical point P, the four key unit vectors are shown in Fig. 5-3. All of them are tangential to the surface at P. A three-dimensional display of these vectors and the normal vector $\mathbf{u}_3$ is shown in Fig. 5-4. In these figures, $\mathbf{u}_l$ is tangential to the edge of the surface, and $\mathbf{u}_m$ is normal to the edge, but tangential to the surface. The relations between these unit vectors are

$$\mathbf{u}_1 \times \mathbf{u}_2 = \mathbf{u}_3 = \mathbf{u}_m \times \mathbf{u}_l. \tag{5.44}$$

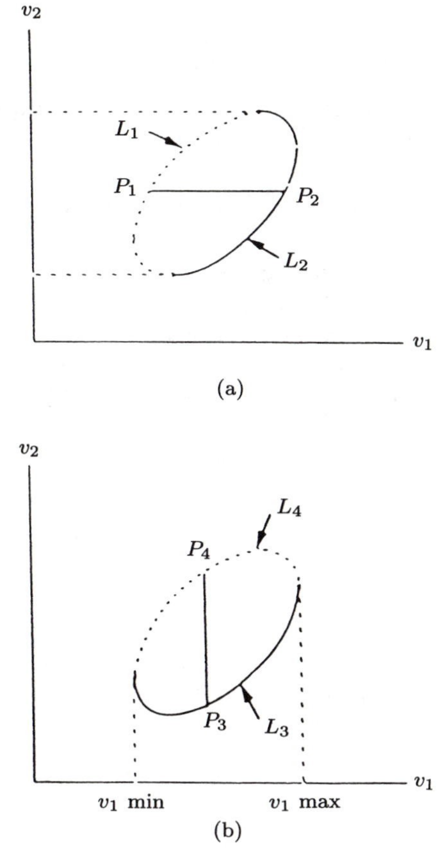

Fig. 5-2 Domain of integration in the (v_1, v_2) plane of a simple region.

The algebraic relations between them are

$$\mathbf{u}_1 = \cos\alpha\mathbf{u}_l + \sin\alpha\mathbf{u}_m \tag{5.45}$$

$$\mathbf{u}_2 = \sin\alpha\mathbf{u}_l - \cos\alpha\mathbf{u}_m \tag{5.46}$$

where α is the angle between $\mathbf{u}_1$ and $\mathbf{u}_l$. If we denote the total differential arc length of the contour at P by dl, then

$$h_1 dv_1 = \cos\alpha dl \tag{5.47}$$

$$h_2 dv_2 = \sin\alpha dl. \tag{5.48}$$

Upon substituting (5.47) and (5.48) into (5.43), we find

$$\mathbf{u}_1 h_2 dv_2 - \mathbf{u}_2 h_1 dv_1 = \mathbf{u}_m dl; \tag{5.49}$$

hence,

$$T(\mathbf{u}_1)h_2 dv_2 - T(\mathbf{u}_2)h_1 dv_1 = T(\mathbf{u}_m)dl. \tag{5.50}$$

Equation (5.39), therefore, reduces to

$$\iint_S T(\nabla_s)dS = \oint_L T(\mathbf{u}_m)dl. \tag{5.51}$$

The unit vector $\mathbf{u}_m$ is commonly denoted by $\mathbf{m}$, in contrast to the notation $\mathbf{n}$ used for $\mathbf{u}_3$; (5.51), therefore, will be written as

$$\iint_S T(\nabla_s)dS = \oint_L T(\mathbf{m})dl. \tag{5.52}$$

Equation (5.52) is designated as the *generalized Gauss theorem for a surface* or the *generalized surface Gauss theorem*. It converts an open surface integral into a closed line integral, and it has the same significance as the generalized Gauss theorem in space which converts a volume integral into a closed surface integral. Various integral theorems can be derived by choosing the proper form for $T(\nabla_s)$.

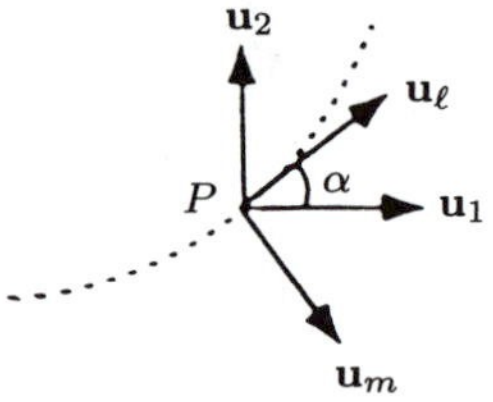

Fig. 5-3 Four tangential vectors in the plane containing $\mathbf{u}_1$ and $\mathbf{u}_2$.

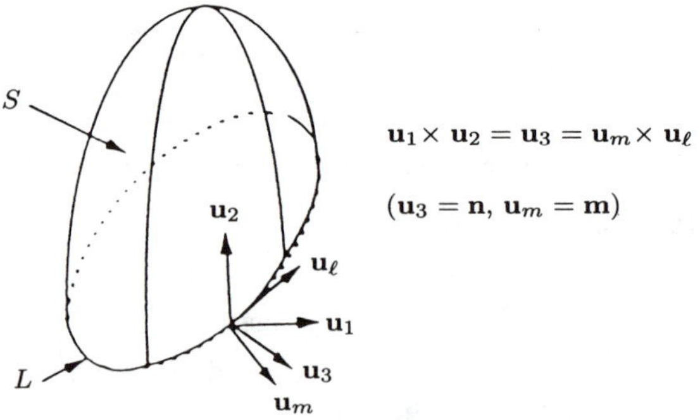

Fig. 5-4 Three-dimensional view of the unit vectors at the edge of an open surface.

1. Surface Gradient Theorem

Let $T(\nabla_s) = \nabla_s f = \nabla_s f$; then $T(\mathbf{m}) = \mathbf{m}f$. Hence,

$$\iint_S \nabla_s f dS = \oint_L f\mathbf{m}dl. \tag{5.53}$$

2. Surface Divergence Theorem

Let $T(\nabla_s) = \nabla_s \cdot \mathbf{f} = \nabla_s \cdot \mathbf{f}$; then $T(\mathbf{m}) = \mathbf{m} \cdot \mathbf{f}$. Hence,

$$\iint_S \nabla_s \cdot \mathbf{f} dS = \oint_L \mathbf{m} \cdot \mathbf{f} dl. \tag{5.54}$$

3. Surface Curl Theorem

Let $T(\nabla_s) = \nabla_s \times \mathbf{f} = \nabla_s \times \mathbf{f}$; then $T(\mathbf{m}) = \mathbf{m} \times \mathbf{f}$. Hence,

$$\iint_S (\nabla_s \times \mathbf{f}) dS = \oint_L (\mathbf{m} \times \mathbf{f}) dl. \tag{5.55}$$

In view of the relationship between $T(\nabla)$ and $T(\nabla_s)$ as described by (5.18), the generalized Gauss theorem for a surface can be written in the form

$$\iint_S \left\{ T(\nabla) - \frac{1}{H} \frac{\partial}{\partial v_3} [HT(\mathbf{u}_3)] \right\} dS = \oint_L T(\mathbf{m})\, d\ell. \tag{5.56}$$

If $T(\nabla)$ is proportional to $\mathbf{u}_3 \times \nabla$ or $\mathbf{n} \times \nabla$, then $T(\mathbf{u}_3) = 0$, and (5.56) becomes

$$\iint_S T(\nabla)\, dS = \oint_L T(\mathbf{m})\, d\ell,\ T(\mathbf{n}) = 0. \tag{5.57}$$

Three cases are considered below.

4. Cross-Gradient Theorem

Let $T(\nabla) = (\mathbf{n} \times \nabla) f$. As a result of Lemmas 1 and 2, we have

$$\begin{aligned} (\mathbf{n} \times \nabla) f &= (\mathbf{n} \times \nabla_n) f + (\mathbf{n} \times \nabla_f) f \\ &= -(\nabla \times \mathbf{n}) f + \mathbf{n} \times \nabla f = \mathbf{n} \times \nabla f \\ T(\mathbf{m}) &= (\mathbf{n} \times \mathbf{m}) f = \mathbf{u}_\ell f. \end{aligned}$$

Hence,

$$\iint_S \mathbf{n} \times \nabla f dS = \oint_L f d\ell. \tag{5.58}$$

For identification purposes, we designate it as the cross-gradient theorem.

5. Stokes Theorem

Let

$$\begin{aligned} T(\nabla) &= (\mathbf{n} \times \nabla) \cdot \mathbf{f} \\ &= (\mathbf{n} \times \nabla_n) \cdot \mathbf{f} + (\mathbf{n} \times \nabla_f) \cdot \mathbf{f} \\ &= -(\nabla \times \mathbf{n}) \cdot \mathbf{f} + \mathbf{n} \cdot (\nabla_f \times \mathbf{f}) = \mathbf{n} \cdot \nabla \times \mathbf{f}. \end{aligned}$$

Then

$$T(\mathbf{m}) = (\mathbf{n} \times \mathbf{m}) \cdot \mathbf{f} = \mathbf{u}_\ell \cdot \mathbf{f}.$$

Hence,

$$\iint_S \mathbf{n} \cdot \nabla \times \mathbf{f} dS = \oint_L \mathbf{f} \cdot d\boldsymbol{\ell}, \tag{5.59}$$

which is the famous theorem named after George Gabriel Stokes (1819–1903).

6. Cross–del–cross Theorem

Let

$$\begin{aligned} T(\nabla\!\!\!\!\nabla) &= (\mathbf{n} \times \nabla\!\!\!\!\nabla) \times \mathbf{f} \\ &= (\mathbf{n} \times \nabla\!\!\!\!\nabla_n) \times \mathbf{f} + (\mathbf{n} \times \nabla\!\!\!\!\nabla_f) \times \mathbf{f} \\ &= -(\nabla \times \mathbf{n}) \times \mathbf{f} + (\mathbf{n} \times \nabla) \times \mathbf{f} = (\mathbf{n} \times \nabla) \times \mathbf{f}. \end{aligned}$$

Then

$$T(\mathbf{m}) = (\mathbf{n} \times \mathbf{m}) \times \mathbf{f} = \mathbf{u}_\ell \times \mathbf{f}.$$

Hence,

$$\iint_S (\mathbf{n} \times \nabla) \times \mathbf{f} dS = -\oint_L \mathbf{f} \times d\boldsymbol{\ell}. \tag{5.60}$$

In Appendix E, it is shown that the Stokes theorem and the cross–del–cross theorem can be derived from the cross–del or cross-gradient theorems.

5-6 SYMBOLIC EXPRESSIONS WITH A SINGLE SYMBOLIC VECTOR AND TWO FUNCTIONS, AND EXPRESSIONS WITH DOUBLE SYMBOLIC VECTORS AND A SINGLE FUNCTION

A complete line of formulas and theorems can be derived covering these two topics. However, we will present only the essential formulation without actually going into detail. A third lemma dealing with symbolic expressions with a single-surface S vector and two functions is one of the main subjects to be covered. A scalar Green's theorem on a surface involving the surface Laplacian will also be presented.

For a symbolic expression with two functions, its differential form in the general Dupin system according to (5.2) is defined by

$$T(\nabla\!\!\!\!\nabla_s, f_1, f_2) = \frac{1}{H} \sum_{i=1}^{2} \frac{\partial}{\partial v_i} \left[\frac{H}{h_i} T(\mathbf{u}_i, f_1, f_2) \right] \tag{5.61}$$

where $H = h_1 h_2$. We now introduce two expressions with two *partial surface S vectors*, denoted by ∇_{s1} and ∇_{s2}, as follows:

$$T(\nabla_{s1}, f_1, f_2) = \frac{1}{H}\sum_{i=1}^{2}\frac{\partial}{\partial v_i}\left[\frac{H}{h_i}T(\mathbf{u}_i, f_1, f_2)\right]_{f_2=c} \tag{5.62}$$

$$T(\nabla_{s2}, f_1, f_2) = \frac{1}{H}\sum_{i=1}^{2}\frac{\partial}{\partial v_i}\left[\frac{H}{h_i}T(\mathbf{u}_i, f_1, f_2)\right]_{f_1=c}. \tag{5.63}$$

It is obvious that Lemma 1 of Chapter 4, Section 4-1 also applies to (5.63) and (5.64). Equation (5.62) can now be decomposed into three parts:

$$\begin{aligned} T(\nabla_s, f_1, f_2) &= T(\nabla_{s1}, f_1, f_2) + T(\nabla_{s2}, f_1, f_2) \\ &\quad - \frac{1}{H}\sum_{i=1}^{2}\frac{\partial}{\partial v_i}\left[\frac{H}{h_i}T(\mathbf{u}_i, f_1, f_2)\right]_{f_1, f_2=c}. \end{aligned} \tag{5.64}$$

This development exactly follows the steps that led us to (4.99). Since $T(\mathbf{u}_i)$ is linear with respect to $\mathbf{u}_i$, we can combine $\mathbf{u}_i$ with Ω/h_i to form one function, and examine its derivatives. Thus,

$$\begin{aligned} \frac{1}{H}\sum_{i=1}^{2}\frac{\partial}{\partial v_i}\left[\frac{H}{h_i}\mathbf{u}_i\right] &= \frac{1}{h_1 h_2}\left[\frac{\partial}{\partial v_1}(h_2\mathbf{u}_1) + \frac{\partial}{\partial v_2}(h_1\mathbf{u}_2)\right] \\ &= \frac{1}{h_1 h_2}\left(h_2\frac{\partial \mathbf{u}_1}{\partial v_1} + h_1\frac{\partial \mathbf{u}_2}{\partial v_2} + \mathbf{u}_1\frac{\partial h_2}{\partial v_1} + \mathbf{u}_2\frac{\partial h_1}{\partial v_2}\right) \\ &= \frac{1}{h_1 h_2}\left[-h_2\left(\frac{\partial h_1}{h_2\partial v_2}\mathbf{u}_2 + \frac{\partial h_1}{\partial v_3}\mathbf{u}_3\right) + \mathbf{u}_1\frac{\partial h_2}{\partial v_1}\right. \\ &\qquad \left. -h_1\left(\frac{\partial h_2}{h_1\partial v_1}\mathbf{u}_1 + \frac{\partial h_2}{\partial v_3}\mathbf{u}_3\right) + \mathbf{u}_2\frac{\partial h_1}{\partial v_2}\right] \\ &= -\left(\frac{1}{h_1}\frac{\partial h_1}{\partial v_3} + \frac{1}{h_2}\frac{\partial h_2}{\partial v_3}\right)\mathbf{u}_3 = J\mathbf{u}_3 \end{aligned} \tag{5.65}$$

where J has been defined by (5.6). Equation (2.26) with $h_3 = 1$ has been used to express the derivatives $\partial\mathbf{u}_i/\partial v_i$ with $i = 1, 2$ in terms of the unit vectors $\mathbf{u}_i$ with $i = 1, 2, 3$. In view of (5.65), the last terms in (5.64) can be written in the form

$$-\frac{1}{H}\sum_{i=1}^{2}\frac{\partial}{\partial v_i}\left[\frac{H}{h_i}T(\mathbf{u}_i, f_1, f_2)\right]_{f_1, f_2=c} = -JT(\mathbf{u}_3, f_1, f_2). \tag{5.66}$$

A lemma, which plays a similar role as Lemma 2 in the vector analysis for functions defined in a three-dimensional space, can now be formulated for functions defined on a two-dimensional curved surface.

Lemma 3: For a symbolic expression defined with respect to a single-surface S vector and two functions, the following relation holds true:

$$T(\boldsymbol{\nabla}_s, f_1, f_2) = T(\boldsymbol{\nabla}_{s1}, f_1, f_2) + T(\boldsymbol{\nabla}_{s2}, f_1, f_2) - JT(\mathbf{u}_3, f_1, f_2). \tag{5.67}$$

The proof of this lemma follows directly from (5.62)–(5.65). Let us illustrate the application of this lemma to the function

$$T(\boldsymbol{\nabla}_s, f_1, f_2) = \boldsymbol{\nabla}_s \cdot (a\mathbf{b}) = \nabla_s \cdot (a\mathbf{b})$$

to obtain

$$\boldsymbol{\nabla}_s \cdot (a\mathbf{b}) = \boldsymbol{\nabla}_{sa} \cdot (a\mathbf{b}) + \boldsymbol{\nabla}_{sb} \cdot (a\mathbf{b}) - J\mathbf{u}_3 \cdot (a\mathbf{b}).$$

Hence,

$$\nabla_s \cdot (a\mathbf{b}) = (\nabla_s a) \cdot \mathbf{b} + a\nabla_s \cdot \mathbf{b} - J\mathbf{u}_3 \cdot (a\mathbf{b}). \tag{5.68}$$

Equation (5.68) has the same significance as identity (4.84); however, there is an extra term as a consequence of Lemma 3. It may be of interest to point out that if we had used Weatherburn's functions for the surface gradient and surface divergence, (5.24) and (5.27), then

$$\nabla_s a = \nabla_t a + J\mathbf{u}_3 a \tag{5.69}$$

$$\nabla_s \cdot \mathbf{b} = \nabla_t \cdot \mathbf{b} + J\mathbf{u}_3 \cdot \mathbf{b} \tag{5.70}$$

$$\nabla_s \cdot (a\mathbf{b}) = \nabla_t \cdot (a\mathbf{b}) + J\mathbf{u}_3 \cdot (a\mathbf{b}). \tag{5.71}$$

Upon substituting them into (5.68), one finds

$$\nabla_t \cdot (a\mathbf{b}) = (\nabla_t a) \cdot \mathbf{b} + a\nabla_t \cdot \mathbf{b}. \tag{5.72}$$

The difference in form between (5.68) and (5.72) is simply due to the different definitions of the two sets of surface functions.

Symbolic expressions involving two surface S vectors and a single function can be evaluated by applying (5.2) repeatedly.

For example, let

$$T(\boldsymbol{\nabla}_s, \boldsymbol{\nabla}_s, f) = \boldsymbol{\nabla}_s \cdot \boldsymbol{\nabla}_s f. \tag{5.73}$$

Then

$$\begin{aligned} \boldsymbol{\nabla}_s \cdot \boldsymbol{\nabla}_s f &= \frac{1}{H} \sum_{i=1}^{2} \frac{\partial}{\partial v_i} \left(\frac{H}{h_i} \frac{1}{h_i} \frac{\partial f}{\partial v_i} \right) \\ &= \frac{1}{H} \sum_{i=1}^{2} \frac{\partial}{\partial v_i} \left(\frac{H}{h_i^2} \frac{\partial f}{\partial v_i} \right), \end{aligned} \tag{5.74}$$

which defines the surface Laplacian of f and is denoted by $\nabla_s^2 f$.

CHAPTER 6

Vector Analysis of Transport Theorems

6-1 HELMHOLTZ TRANSPORT THEOREM

Thus far, we have been dealing only with functions of position, i.e., functions that are dependent on spatial variables only. In many engineering and physical problems, the quantities involved are functions of both space and time. Examples are the induced voltage in a moving coil of an electric generator and the transport of fluid in a channel. The mathematical formulation of these problems requires a knowledge of vector analysis involving a moving surface or a moving body. One of the fundamental theorems in this area is the Helmholtz transport theorem, named after the renowned German scientist Hermann Ludwig Ferdinand von Helmholtz (1821–1894). The theorem deals with the time rate of change of a surface integral of Type IV stated by (3.70), in which the domain of integration and the integrand are functions of both space and time. The quantity under consideration is defined by

$$\begin{aligned} I &= \frac{d}{dt} \iint_{S\,(\mathbf{R},t)} \mathbf{F}\,(\mathbf{R},t) \cdot d\mathbf{S} \\ &= \lim_{\Delta t \to 0} \frac{1}{\Delta t} \left[\iint_{S_2\,(\mathbf{R},t+\Delta t)} \mathbf{F}\,(\mathbf{R},t+\Delta t) \cdot d\mathbf{S} \right. \\ &\quad \left. - \iint_{S_1\,(\mathbf{R},t)} \mathbf{F}\,(\mathbf{R},t) \cdot d\mathbf{S} \right] \end{aligned} \tag{6.1}$$

where $\mathbf{F}(\mathbf{R},t)$ is an abbreviated notation for $\mathbf{F}(x_1,x_2,x_3,t)$. The x_i's denote the coordinate variables of the position vector where the function $\mathbf{F}$ is defined, and t is the time variable. The domain of integration changes from $S_1(\mathbf{R},t)$ to $S_2(\mathbf{R},t+\Delta t)$ in a small time interval Δt, as shown in Fig. 6-1. To evaluate the limiting value of the difference of the two surface integrals contained in (6.1), we first expand the integrand $\mathbf{F}(\mathbf{R},t+\Delta t)$ in a Taylor series with respect to t:

$$\mathbf{F}(\mathbf{R},t+\Delta t) = \mathbf{F}(\mathbf{R},t) + \frac{\partial \mathbf{F}(\mathbf{R},t)}{\partial t}\Delta t + \frac{1}{2}\frac{\partial^2 \mathbf{F}(\mathbf{R},t)}{\partial t^2}(\Delta t)^2 + \cdots. \tag{6.2}$$

Upon substituting (6.2) into (6.1), we have

$$\begin{aligned} I &= \lim_{\Delta t\to 0}\frac{1}{\Delta t}\left\{\iint_{S_2(\mathbf{R},t+\Delta t)}\left[\mathbf{F}(\mathbf{R},t)+\frac{\partial \mathbf{F}(\mathbf{R},t)}{\partial t}\Delta t+\cdots\right]\cdot d\mathbf{S} \right.\\ &\quad \left. - \iint_{S_1(\mathbf{R},t)}\mathbf{F}(\mathbf{R},t)\cdot d\mathbf{S}\right\} \\ &= \iint_{S(\mathbf{R},t)}\frac{\partial \mathbf{F}(\mathbf{R},t)}{\partial t}\cdot d\mathbf{S} + \lim_{\Delta t\to 0}\frac{1}{\Delta t}\left[\iint_{S_2(\mathbf{R},t+\Delta t)}\mathbf{F}(\mathbf{R},t)\cdot d\mathbf{S}\right. \\ &\quad \left. - \iint_{S_1(\mathbf{R},t)}\mathbf{F}(\mathbf{R},t)\cdot d\mathbf{S}\right]. \end{aligned} \tag{6.3}$$

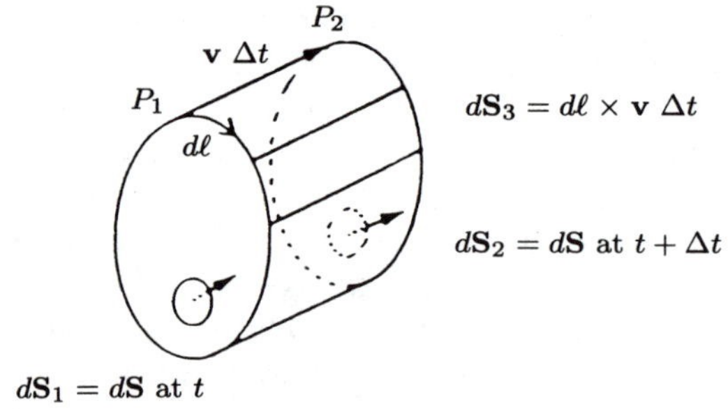

Fig. 6-1 Moving surface at two different instants.

In (6.3), $S_1(\mathbf{R},t)$ is the same as $S(\mathbf{R},t)$; the subscripts "1" and "2" are used to identify the location of the surface at time t and at a later time $t+\Delta t$, respectively, as shown in Fig. 6-1. A point P_1 at the contour of S_1 is displaced to a point P_2 at the contour of S_2 during the time interval Δt. The displacement is equal to $\mathbf{v}\Delta t$ where $\mathbf{v}$ denotes the velocity of motion at that location, which may vary continuously from one location to another around the contour. For example, when a circular loop spins around its diagonal axis, the linear velocity varies around its circumference. The two surface integrals within the brackets of (6.3) can be written in the form

$$\iint_{S_2(\mathbf{R},t+\Delta t)} \mathbf{F}(\mathbf{R},t)\cdot d\mathbf{S} - \iint_{S_1(\mathbf{R},t)} \mathbf{F}(\mathbf{R},t)\cdot d\mathbf{S}$$

$$= \oiint_{S_1+S_2+S_3} \mathbf{F}(\mathbf{R},t)\cdot d\mathbf{S} - \iint_{S_3} \mathbf{F}(\mathbf{R},t)\cdot d\mathbf{S}_3 \qquad (6.4)$$

where S_3 denotes the lateral surface swept by the displacement vector $\mathbf{v}\Delta t$ as S_1 moves to S_2. It is observed that $d\mathbf{S}_1$ is pointed into the volume bounded by S_1, S_2, and S_3, while $d\mathbf{S}_2$ is pointed outward. The closed surface integral in (6.4) can be transformed to a volume integral, i.e.,

$$\oiint_S \mathbf{F}(\mathbf{R},t)\cdot d\mathbf{S} = \iiint_V \nabla\cdot\mathbf{F}(\mathbf{R},t)\,dV. \qquad (6.5)$$

In (6.4) and (6.5),

$$d\mathbf{S}_3 = d\boldsymbol{\ell}\times(\mathbf{v}\Delta t) \qquad (6.6)$$

$$dV = (\mathbf{v}\Delta t)\cdot d\mathbf{S}. \qquad (6.7)$$

By making use of the mean-value theorem in calculus, the surface integral in (6.4) evaluated on S_3 and the volume integral in (6.5) can be written in the following form:

$$-\iint_{S_3} \mathbf{F}(\mathbf{R},t)\cdot d\mathbf{S}_3 = \Delta t \oint_L \mathbf{F}(\mathbf{R},t)\cdot(\mathbf{v}\times d\boldsymbol{\ell})$$

$$= -\Delta t \oint_L [\mathbf{v}\times\mathbf{F}(\mathbf{R},t)]\cdot d\boldsymbol{\ell} \qquad (6.8)$$

$$\iiint_V \nabla\cdot\mathbf{F}(\mathbf{R},t)\,dV = \Delta t \iint_S [\mathbf{v}\,\nabla\cdot\mathbf{F}(\mathbf{R},t)]\cdot d\mathbf{S}. \qquad (6.9)$$

Equation (6.4) now becomes

$$\iint_{S_2(\mathbf{R},t+\Delta t)} \mathbf{F}(\mathbf{R},t)\cdot d\mathbf{S} - \iint_{S_1(\mathbf{R},t)} \mathbf{F}(\mathbf{R},t)\cdot d\mathbf{S}$$

$$= \Delta t\left\{\iint_S [\mathbf{v}\,\nabla\cdot\mathbf{F}(\mathbf{R},t)]\cdot d\mathbf{S} - \oint_L [\mathbf{v}\times\mathbf{F}(\mathbf{R},t)]\cdot d\boldsymbol{\ell}\right\}$$

$$= \Delta t\left\{\iint_S \{\mathbf{v}\,\nabla\cdot\mathbf{F}(\mathbf{R},t) - \nabla\times[\mathbf{v}\times\mathbf{F}(\mathbf{R},t)]\}\cdot d\mathbf{S}\right\}. \qquad (6.10)$$

In (6.10), the line integral has been converted to a surface integral by means of the Stokes theorem. Equation (6.3), after taking the limit with respect to Δt, yields

$$\frac{d}{dt}\iint_{S(\mathbf{R},t)} \mathbf{F}(\mathbf{R},t)\cdot d\mathbf{S} = \iint_{S(\mathbf{R},t)} \left\{ \frac{\partial \mathbf{F}(\mathbf{R},t)}{\partial t} + \mathbf{v}\,\nabla\cdot\mathbf{F}(\mathbf{R},t) - \nabla\times[\mathbf{v}\times\mathbf{F}(\mathbf{R},t)] \right\}\cdot d\mathbf{S}$$

or simply

$$\frac{d}{dt}\iint_S \mathbf{F}\cdot d\mathbf{S} = \iint_S \left[\frac{\partial \mathbf{F}}{\partial t} + \mathbf{v}\,\nabla\cdot\mathbf{F} - \nabla\times(\mathbf{v}\times\mathbf{F})\right]\cdot d\mathbf{S}, \tag{6.11}$$

which represents the Helmholtz transport theorem [7] using the modern notation of vector analysis first formulated by Lorentz [8], and reiterated by Sommerfeld [13]. Equation (6.11) can be cast in a different form by making use of identity (4.92) for $\nabla\times(\mathbf{v}\times\mathbf{F})$, which yields

$$\frac{d}{dt}\iint_S \mathbf{F}\cdot d\mathbf{S} = \iint_S \left[\frac{\partial \mathbf{F}}{\partial t} + (\mathbf{v}\cdot\nabla)\,\mathbf{F} + \mathbf{F}\nabla\cdot\mathbf{v} - (\mathbf{F}\cdot\nabla)\,\mathbf{v}\right]\cdot d\mathbf{S}. \tag{6.12}$$

This version of the Helmholtz transport theorem is used by Candel and Poinsot [2] in formulating a problem in gas dynamics.

Since the material derivative of $\mathbf{F}$ or the total time derivative of $\mathbf{F}$ is defined by

$$\frac{d\mathbf{F}}{dt} = \frac{\partial \mathbf{F}}{\partial t} + \sum_{i=1}^{3} \frac{\partial \mathbf{F}}{\partial x_i}\frac{\partial x_i}{\partial t} = \frac{\partial \mathbf{F}}{\partial t} + (\mathbf{v}\cdot\nabla)\,\mathbf{F}, \tag{6.13}$$

(6.12) can be written in the form

$$\frac{d}{dt}\iint_S \mathbf{F}\cdot d\mathbf{S} = \iint_S \left[\frac{d\mathbf{F}}{dt} + \mathbf{F}\nabla\cdot\mathbf{v} - (\mathbf{F}\cdot\nabla)\,\mathbf{v}\right]\cdot d\mathbf{S}. \tag{6.14}$$

This form of the Helmholtz theorem is found in the treatment by Truesdell and Toupin [19].

6-2 MAXWELL THEOREM AND REYNOLDS TRANSPORT THEOREM

Two related theorems can now be derived from the Helmholtz transport theorem, although in the original works, these two theorems were formulated independently of the Helmholtz theorem. In the Helmholtz theorem, if we let

$$\mathbf{F} = \nabla\times\mathbf{f} \tag{6.15}$$

and then convert the surface integral into a line integral, we find, noting that $\nabla\cdot\mathbf{F} = 0$ in view of (4.121),

$$\frac{d}{dt}\oint_S \mathbf{f}\cdot d\boldsymbol{\ell} = \oint_L \left(\frac{\partial \mathbf{f}}{\partial t} - \mathbf{v}\times\nabla\times\mathbf{f}\right)\cdot d\boldsymbol{\ell}. \tag{6.16}$$

This is the statement of the Maxwell theorem originally found in his great work on electromagnetic theory [9], [15].

In the Helmholtz theorem, if the surface is a closed one, we obtain

$$\frac{d}{dt}\oiint \mathbf{F}\cdot d\mathbf{S} = \oiint \left(\frac{\partial \mathbf{F}}{\partial t} + \mathbf{v}\,\nabla\cdot\mathbf{F}\right)\cdot d\mathbf{S}. \tag{6.17}$$

The closed surface integral of $\nabla \times (\mathbf{v} \times \mathbf{F})$ vanishes because it is equal to a volume integral of $\nabla \cdot [\nabla \times (\mathbf{v} \times \mathbf{F})]$ which vanishes identically because of (4.121).

As a consequence of the Gauss theorem, (6.17) can be changed into the form

$$\frac{d}{dt}\iiint \nabla\cdot\mathbf{F}dV = \iiint \left[\frac{\partial}{\partial t}\nabla\cdot\mathbf{F} + \nabla\cdot(\mathbf{v}\nabla\cdot\mathbf{F})\right] dV. \tag{6.18}$$

Now, if we let $\nabla\cdot\mathbf{F} = \rho$, a scalar function, then we obtain the Reynolds transport theorem [12], namely,

$$\frac{d}{dt}\iiint \rho dV = \iiint \left[\frac{\partial \rho}{\partial t} + \nabla\cdot(\rho\mathbf{v})\right] dV \tag{6.19}$$

where we identify $\mathbf{v}$ as the velocity of the fluid with density ρ. Since

$$\nabla\cdot(\rho\mathbf{v}) = \rho\nabla\cdot\mathbf{v} + \mathbf{v}\cdot\nabla\rho \tag{6.20}$$

and the total time derivative, or the material derivative, of ρ is given by

$$\frac{d\rho}{dt} = \frac{\partial\rho}{\partial t} + \mathbf{v}\cdot\nabla\rho,$$

(6.19) can be written in the form

$$\frac{d}{dt}\iiint \rho dV = \iiint \left[\frac{d\rho}{dt} + \rho\nabla\cdot\mathbf{v}\right] dV. \tag{6.21}$$

It should be mentioned that the original work of Reynolds (6.19) was derived by evaluating the total time derivative of $\iiint \rho dV$:

$$\begin{aligned}\frac{d}{dt}\iiint \rho dV &= \lim_{\Delta t\to 0}\frac{1}{\Delta t}\left[\iiint_{V(t+\Delta t)} \rho(t+\Delta t)dV\right.\\ &\quad \left. - \iiint_{V(t)} \rho(t)dV\right]\end{aligned}$$

in a manner very similar to the derivation of the Helmholtz theorem.

From the preceding discussion it is seen that the Helmholtz transport theorem can be considered as the principal transport theorem; both the Maxwell theorem and the Reynolds theorem can be treated as lemmas of that theorem.

CHAPTER 7

Dyadic Analysis

7-1 DYADIC ALGEBRA

In the hands of John Willard Gibbs (1839–1903), an American physicist renowned for his work in applied mathematics and thermodynamics, the modern notation of vector analysis was firmly established. A vector function expressed in a Cartesian coordinate system is written in the form

$$\mathbf{F} = \sum_i F_i \mathbf{a}_i. \tag{7.1}$$

The summation index "i" is always carried out from 1 to 3 (in this chapter). He extended this form to define a *dyadic function* or a *dyadic* for short, denoted by [5]

$$\bar{\bar{\mathrm{F}}} = \sum_j \mathbf{F}_j \mathbf{a}_j \tag{7.2}$$

where

$$\mathbf{F}_j = \sum_i F_{ij} \mathbf{a}_i, \qquad i = 1, 2, 3 \tag{7.3}$$

represents three independent or distinct vector functions. The positioning of $\mathbf{F}_j$ and $\mathbf{a}_j$ is kept in that order, and we are not supposed to interchange the ordering of these two vectors. In other words, the commutative rule does not apply to (7.2). When (7.3) is substituted into (7.2), we obtain

$$\bar{\bar{\mathrm{F}}} = \sum_i \sum_j F_{ij} \mathbf{a}_i \mathbf{a}_j. \tag{7.4}$$

We have enforced the distributive rule in changing (7.2) to (7.4). Equations (7.2)–(7.4) contain the definition of a dyadic in a Cartesian coordinate system. There are nine scalar components of $\bar{\bar{\mathrm{F}}}$. The double unit vectors, juxtaposed together, $\mathbf{a}_i$ with $\mathbf{a}_j$ with $i, j = (1, 2, 3)$, are designated as *dyads*, and there are nine of them too. A typical term like $F_{23}\mathbf{a}_2\mathbf{a}_3$ is referred to as a dyadic component. The dyads are not commutative, i.e.,

$$\mathbf{a}_i\mathbf{a}_j \neq \mathbf{a}_j\mathbf{a}_i. \tag{7.5}$$

Transpose

The *transpose* of a dyadic $\bar{\bar{\mathrm{F}}}$, denoted by $[\bar{\bar{\mathrm{F}}}]^T$, is defined by

$$[\bar{\bar{\mathrm{F}}}]^T = \sum_j \mathbf{a}_j\mathbf{F}_j = \sum_i\sum_j F_{ij}\mathbf{a}_j\mathbf{a}_i = \sum_i\sum_j F_{ji}\mathbf{a}_i\mathbf{a}_j. \tag{7.6}$$

In comparing (7.6) to (7.4), we see that the positioning of $\mathbf{a}_i$ and $\mathbf{a}_j$ in the third term of (7.6) has been interchanged or the function F_{ij} in (7.4) is replaced by F_{ji} in the last term of (7.6).

Symmetrical and Antisymmetrical Dyadics

A symmetrical dyadic, denoted by $\bar{\bar{\mathrm{F}}}_s$, is characterized by

$$[\bar{\bar{\mathrm{F}}}_s]^T = \bar{\bar{\mathrm{F}}}_s. \tag{7.7}$$

A symmetrical dyadic, therefore, has only six distinct scalar components because $F_{ij} = F_{ji}$.

An antisymmetrical dyadic, denoted by $\bar{\bar{\mathrm{F}}}_a$, is characterized by

$$[\bar{\bar{\mathrm{F}}}_a]^T = -\bar{\bar{\mathrm{F}}}_a. \tag{7.8}$$

Thus,

$$F_{ij} = -F_{ji}. \tag{7.9}$$

Equation (7.9) implies that

$$F_{ii} = 0, \qquad i = 1, 2, 3. \tag{7.10}$$

$\bar{\bar{\mathrm{F}}}_a$, therefore, has only three distinctive scalar components if we do not consider the negative sign in (7.9) as distinctly different. One special case of a symmetrical dyadic is when $F_{ij} = 0$ for $i \neq j$, and $F_{ii} = 1$ for $i = 1, 2, 3$. These two properties can be expressed by

$$F_{ij} = \delta_{ij} = \begin{cases} 1, & i = j \\ 0, & i \neq j \end{cases} \tag{7.11}$$

where δ_{ij} denotes the Kronecker delta function. This dyadic is called an idem factor or idem–dyad, and is denoted by $\bar{\bar{\mathrm{I}}}$. Its explicit expression is

$$\bar{\bar{\mathrm{I}}} = \sum_i \mathbf{a}_i\mathbf{a}_i = \mathbf{u}_x\mathbf{u}_x + \mathbf{u}_y\mathbf{u}_y + \mathbf{u}_z\mathbf{u}_z. \tag{7.12}$$

Scalar Products of a Vector and a Dyadic

There are two scalar products between a vector and a dyadic. The anterior scalar product, denoted by $\mathbf{b} \cdot \bar{\bar{\mathrm{F}}}$, is defined by

$$\mathbf{b} \cdot \bar{\bar{\mathrm{F}}} = \sum_j (\mathbf{b} \cdot \mathbf{F}_j)\mathbf{a}_j = \sum_i \sum_j (\mathbf{b} \cdot F_{ij}\mathbf{a}_i)\mathbf{a}_j$$

$$= \sum_i \sum_j (b_i F_{ij})\mathbf{a}_j, \tag{7.13}$$

which is a vector. If we denote this vector by $\mathbf{c}$, then

$$c_j = \sum_i b_i F_{ij}, \qquad j = 1, 2, 3. \tag{7.14}$$

The posterior scalar product of a vector and a dyadic, denoted by $\bar{\bar{\mathrm{F}}} \cdot \mathbf{b}$, is defined by

$$\bar{\bar{\mathrm{F}}} \cdot \mathbf{b} = \sum_j \mathbf{F}_j(\mathbf{a}_j \cdot \mathbf{b}) = \sum_i \sum_j F_{ij}\mathbf{a}_i(\mathbf{a}_j \cdot \mathbf{b})$$

$$= \sum_i \sum_j (b_j F_{ij})\mathbf{a}_i, \tag{7.15}$$

which is also a vector. If we denote this vector by $\mathbf{d}$, then

$$d_i = \sum_j b_j F_{ij}, \qquad j = 1, 2, 3. \tag{7.16}$$

By comparing (7.14) to (7.16), we see that, in general, $c_j \neq d_j$, except for a symmetrical dyadic, which implies that

$$\mathbf{b} \cdot \bar{\bar{\mathrm{F}}}_s = \bar{\bar{\mathrm{F}}}_s \cdot \mathbf{b}. \tag{7.17}$$

In particular,

$$\mathbf{b} \cdot \bar{\bar{\mathrm{I}}} = \bar{\bar{\mathrm{I}}} \cdot \mathbf{b} = \mathbf{b}. \tag{7.18}$$

This is an important characteristic of the idem factor. Based on the definition of the transpose of a dyadic (7.6) and (7.13), it is obvious that

$$\mathbf{b} \cdot \bar{\bar{\mathrm{F}}} = [\bar{\bar{\mathrm{F}}}]^T \cdot \mathbf{b} \tag{7.19}$$

or

$$\bar{\bar{\mathrm{F}}} \cdot \mathbf{c} = \mathbf{c} \cdot [\bar{\bar{\mathrm{F}}}]^T. \tag{7.20}$$

This relation is frequently used in dyadic analysis.

Vector Products of a Vector and a Dyadic

There are again two vector products between a vector and a dyadic. The anterior vector product, denoted by $\mathbf{b} \times \bar{\bar{\mathrm{F}}}$, is defined by

$$\mathbf{b} \times \bar{\bar{\mathrm{F}}} = \sum_j (\mathbf{b} \times \mathbf{F}_j)\mathbf{a}_j = \sum_i \sum_j F_{ij}(\mathbf{b} \times \mathbf{a}_i)\mathbf{a}_j, \tag{7.21}$$

which is a dyadic. The posterior vector product of a vector and a dyadic, denoted by $\bar{\bar{\mathrm{F}}} \times \mathbf{b}$, is defined by

$$\bar{\bar{\mathrm{F}}} \times \mathbf{b} = \sum_j \mathbf{F}_j(\mathbf{a}_j \times \mathbf{b}) = \sum_i \sum_j F_{ij}\mathbf{a}_i(\mathbf{a}_j \times \mathbf{b}), \tag{7.22}$$

which is another dyadic. In general,

$$\mathbf{b} \times \bar{\bar{\mathrm{F}}} \neq -\bar{\bar{\mathrm{F}}} \times \mathbf{b} \tag{7.23}$$

and

$$\mathbf{b} \times \bar{\bar{\mathrm{F}}} \neq [\bar{\bar{\mathrm{F}}}]^T \times \mathbf{b}. \tag{7.24}$$

It can be verified that

$$\mathbf{b} \times \bar{\bar{\mathrm{F}}} = -\left\{[\bar{\bar{\mathrm{F}}}]^T \times \mathbf{b}\right\}^T. \tag{7.25}$$

In the special case,

$$\mathbf{b} \times \bar{\bar{\mathrm{I}}} = -[\bar{\bar{\mathrm{I}}} \times \mathbf{b}]^T.$$

Triple Products

In vector algebra, we have the identity

$$\mathbf{b} \cdot (\mathbf{c} \times \mathbf{d}) = \mathbf{d} \cdot (\mathbf{b} \times \mathbf{c}) = \mathbf{c} \cdot (\mathbf{d} \times \mathbf{b}). \tag{7.26}$$

(We have avoided the use of $\mathbf{a}$ as a vector in this section because of the conflict of notation with the unit vector $\mathbf{a}_i$ now being used.) A similar identity can be generated for dyadics based on (7.26). We consider three sets of this kind of identity involving three distinct vectors $\mathbf{d}_j$ with $j = 1, 2, 3$, i.e.,

$$\mathbf{b} \cdot (\mathbf{c} \times \mathbf{d}_j) = (\mathbf{b} \times \mathbf{c}) \cdot \mathbf{d}_j = -\mathbf{c} \cdot (\mathbf{b} \times \mathbf{d}_j), \qquad j = 1, 2, 3. \tag{7.27}$$

There is a good reason for us to put the vector $\mathbf{d}_j$ at the posterior position in (7.27), as will be revealed immediately. Now, we juxtapose a unit vector $\mathbf{a}_j$ at the posterior position of each term in (7.27), and take the sum of these three dyadic expressions, which yields

$$\mathbf{b} \cdot (\mathbf{c} \times \bar{\bar{\mathrm{d}}}) = (\mathbf{b} \times \mathbf{c}) \cdot \bar{\bar{\mathrm{d}}} = -\mathbf{c} \cdot (\mathbf{b} \times \bar{\bar{\mathrm{d}}}). \tag{7.28}$$

Each term in (7.28) is a vector because the scalar product of a vector and a dyadic is a vector. Thus, we have elevated the formula for the vector triple products to a level involving one dyadic such as $\mathbf{c} \times \bar{\bar{\mathrm{d}}}$, $\bar{\bar{\mathrm{d}}}$, and $\mathbf{b} \times \bar{\bar{\mathrm{d}}}$, and a vector, that is, $\mathbf{b}$, $\mathbf{b} \times \mathbf{c}$, and $-\mathbf{c}$ in (7.28). We can further elevate the vector $\mathbf{c}$ into a dyadic in the last two terms of (7.28) by considering three distinct equations of the form

$$[\bar{\bar{\mathrm{d}}}]^T \cdot (\mathbf{b} \times \mathbf{c}_j) = -[\mathbf{b} \times \bar{\bar{\mathrm{d}}}]^T \cdot \mathbf{c}_j, \qquad j = 1, 2, 3 \tag{7.29}$$

where we have applied (7.20) to the last two terms of (7.28) to do the conversion. By summing the two dyadic products after juxtaposing $\mathbf{a}_j$ at the posterior position in (7.29), we obtain the following identity:

$$[\bar{\bar{\mathrm{d}}}]^T \cdot (\mathbf{b} \times \bar{\bar{\mathrm{c}}}) = -[\mathbf{b} \times \bar{\bar{\mathrm{d}}}]^T \cdot \bar{\bar{\mathrm{c}}}. \tag{7.30}$$

Each member is now a dyadic, so it is truly a dyadic identity. We have thus completed a basic exposure of the elements of dyadic algebra with all of the vectors and the dyadics expressed in a Cartesian coordinate system.

In the application of dyadic analysis to the formulation of many physical problems, we will encounter dyadics which are defined with respect to two vector functions in two *independent* coordinate systems. In the general orthogonal system, a vector function which is a function of (v_1, v_2, v_3) and $(\mathbf{u}_1, \mathbf{u}_2, \mathbf{u}_3)$ will be written in the form

$$\mathbf{M}(\mathbf{R}) = \sum_i M_i(\mathbf{R})\mathbf{u}_i \tag{7.31}$$

where $\mathbf{R}$ denotes the position vector of a point P located at (v_1, v_2, v_3) of that system. Another vector function that depends on (v_1', v_2', v_3') and $(\mathbf{u'}_1, \mathbf{u'}_2, \mathbf{u'}_3)$ will be written in the form

$$\mathbf{N}'(\mathbf{R}') = \sum_j N_j(\mathbf{R}')\mathbf{u'}_j \tag{7.32}$$

where $\mathbf{R}'$ denotes the position vector of another point P' located at (v_1', v_2', v_3') of the primed system. Figure 7-1 shows the position vectors $\mathbf{R}$ and $\mathbf{R}'$ and the coordinate systems associated with the two vector functions. The independence of the two systems means that

$$\frac{\partial \mathbf{M}(\mathbf{R})}{\partial v_i'} = 0 \ \text{ and } \ \frac{\partial \mathbf{N}'(\mathbf{R}')}{\partial v_i} = 0. \tag{7.33}$$

A dyadic formed by these two vectors, denoted by $\bar{\bar{\mathrm{D}}}$, is defined by

$$\begin{aligned} \bar{\bar{\mathrm{D}}} &= \mathbf{M}(\mathbf{R})\mathbf{N}'(\mathbf{R}') \\ &= \left(\sum_i M_i(\mathbf{R})\mathbf{u}_i\right)\left(\sum_j N_j(\mathbf{R}')\mathbf{u'}_j\right) \\ &= \sum_i \sum_j M_i N_j' \mathbf{u}_i \mathbf{u'}_j \end{aligned} \tag{7.34}$$

where M_i is a function of (v_1, v_2, v_3) and N_j' is a function of (v_1', v_2', v_3'). The transpose of $\bar{\bar{\mathrm{D}}}$ is given by

$$\begin{aligned} [\bar{\bar{\mathrm{D}}}]^T &= \sum_i \sum_j M_i N_j' \mathbf{u'}_j \mathbf{u}_i \\ &= \mathbf{N}'(\mathbf{R}')\mathbf{M}(\mathbf{R}). \end{aligned} \tag{7.35}$$

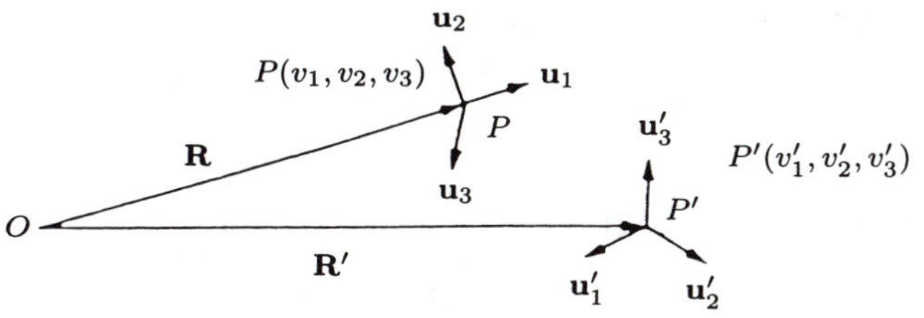

Fig. 7-1 Two independent coordinate systems associated with two independent position vectors.

The differential calculus or the dyadic analysis of these dyadic functions are discussed in the following section.

7-2 DIVERGENCE AND CURL OF DYADIC FUNCTIONS AND GRADIENT OF VECTOR FUNCTIONS

The divergence of a dyadic function $\bar{\bar{\mathrm{F}}}$ expressed in a Cartesian system by (7.2)–(7.4) in the previous section, denoted by $\nabla \cdot \bar{\bar{\mathrm{F}}}$, is defined by

$$\nabla \cdot \bar{\bar{\mathrm{F}}} = \sum_j (\nabla \cdot \mathbf{F}_j)\mathbf{a}_j = \sum_i \sum_j \frac{\partial F_{ij}}{\partial x_i}\mathbf{a}_j, \tag{7.36}$$

which is a vector function. If the dyadic function is defined in two independent orthogonal systems in the form of (7.34), then its definition is

$$\nabla \cdot \bar{\bar{\mathrm{D}}} = (\nabla \cdot \mathbf{M})\mathbf{N}' = \left[\sum_i \frac{1}{\Omega}\frac{\partial}{\partial v_i}\left(\frac{\Omega}{h_i}M_i\right)\right]\mathbf{N}' \tag{7.37}$$

and the divergence of $[\bar{\bar{\mathrm{D}}}]^T$ in the primed system is defined by

$$\nabla' \cdot [\bar{\bar{\mathrm{D}}}]^T = (\nabla' \cdot \mathbf{N}')\mathbf{M} = \left[\sum_j \frac{1}{\Omega'}\frac{\partial}{\partial v_j'}\left(\frac{\Omega'}{h_i'}N_j'\right)\right]\mathbf{M} \tag{7.38}$$

where ∇' denotes the del operator in the primed coordinate system. As a result of these definitions,

$$\nabla' \cdot \bar{\bar{\mathrm{D}}} = 0 \ \text{ and } \ \nabla \cdot [\bar{\bar{\mathrm{D}}}]^T = 0. \tag{7.39}$$

The curl of a dyadic function of the form $\bar{\bar{\mathrm{F}}}$, denoted by $\nabla \times \bar{\bar{\mathrm{F}}}$, is defined by

$$\nabla \times \bar{\bar{\mathrm{F}}} = \sum_j (\nabla \times \mathbf{F}_j)\mathbf{a}_j = \sum_i \sum_j (\nabla F_{ij} \times \mathbf{a}_i)\mathbf{a}_j, \tag{7.40}$$

which is also a dyadic function. Here, we have used the vector identity

$$\nabla \times \mathbf{F}_j = \nabla \times \sum_i F_{ij}\mathbf{a}_i = \sum_i (\nabla F_{ij} \times \mathbf{a}_i) \tag{7.41}$$

to convert the single sum in (7.40) to a double sum. The curl of a dyadic in the form of $\bar{\bar{\mathrm{D}}}$ given by (7.34), denoted by $\nabla \times \bar{\bar{\mathrm{D}}}$, is defined by

$$\nabla \times \bar{\bar{\mathrm{D}}} = (\nabla \times \mathbf{M})\mathbf{N}' \tag{7.42}$$

where $\nabla \times \mathbf{M}$ is given by (4.42) in the general orthogonal system. Similarly,

$$\nabla' \times [\bar{\bar{\mathrm{D}}}]^T = (\nabla' \times \mathbf{N}')\mathbf{M}. \tag{7.43}$$

By definition,

$$\nabla' \times \bar{\bar{\mathrm{D}}} = 0 \;\text{ and }\; \nabla \times [\bar{\bar{\mathrm{D}}}]^T = 0. \tag{7.44}$$

Sometimes, we need the gradient of a vector function in dyadic analysis, denoted by $\nabla\mathbf{f}$. In a general orthogonal system, it is defined by

$$\nabla\mathbf{f} = \nabla \sum_j f_j \mathbf{u}_j = \sum_i \sum_j \frac{\mathbf{u}_i}{h_i} \frac{\partial}{\partial v_i}(f_j \mathbf{u}_j). \tag{7.45}$$

The function so defined is obviously a dyadic function. The anterior scalar product of $\nabla\mathbf{f}$ and a vector $\mathbf{g}$ will be

$$\mathbf{g} \cdot \nabla\mathbf{f} = \sum_i \sum_j \frac{g_i}{h_i} \frac{\partial}{\partial v_i}(f_j \mathbf{u}_j). \tag{7.46}$$

In (4.104), this vector function is written in the form of $(\mathbf{g} \cdot \nabla)\mathbf{f}$ with $\mathbf{a}$ replaced by $\mathbf{g}$ and $\mathbf{b}$ replaced by $\mathbf{f}$. In dyadic notation, the brackets are not needed.

When a dyadic function is formed by a scalar function f with an idem factor $\bar{\bar{\mathrm{I}}}$ in the form of

$$f\bar{\bar{\mathrm{I}}} = \sum_i f\mathbf{a}_i\mathbf{a}_i, \tag{7.47}$$

the divergence of this dyadic function is then a vector function, and it is given by

$$\nabla \cdot (f\bar{\bar{\mathrm{I}}}) = \sum_i \frac{\partial f}{\partial x_i}\mathbf{a}_i = \nabla f, \tag{7.48}$$

which is a vector. In a general orthogonal system, $\bar{\bar{\mathrm{I}}}$ is defined by

$$\bar{\bar{\mathrm{I}}} = \sum_i \mathbf{u}_i\mathbf{u}_i \tag{7.49}$$

and the divergence of $f\bar{\bar{\mathrm{I}}}$ is defined by

$$\begin{aligned}
\nabla \cdot (f\bar{\bar{\mathrm{I}}}) &= \sum_i \left[\frac{\mathbf{u}_i}{h_i} \cdot \frac{\partial}{\partial v_i}(f\mathbf{u}_i)\right]\mathbf{u}_i \\
&= \sum_i \frac{\mathbf{u}_i}{h_i} \cdot \left[f\frac{\partial \mathbf{u}_i}{\partial v_i} + \mathbf{u}_i\frac{\partial f}{\partial v_i}\right]\mathbf{u}_i.
\end{aligned} \tag{7.50}$$

In the above equation,

$$\mathbf{u}_i \cdot \frac{\partial \mathbf{u}_i}{\partial v_i} = 0, \tag{7.51}$$

which can be proved by taking the derivative of

$$\mathbf{u}_i \cdot \mathbf{u}_i = 1$$

with respect to v_i or any variable for that matter. It is also known that $\partial \mathbf{u}_i / \partial v_i$ can be expressed in terms of two unit vectors perpendicular to $\mathbf{u}_i$, (2.26). Thus, we obtain

$$\nabla \cdot (f \bar{\bar{\mathbf{I}}}) = \sum_i \frac{1}{h_i} \frac{\partial f}{\partial v_i} \mathbf{u}_i = \nabla f. \tag{7.52}$$

Equation (7.52), therefore, is invariant to the coordinate system. By following the same approach, we find

$$\nabla \times (f \bar{\bar{\mathbf{I}}}) = \sum_j (\nabla \times f \mathbf{a}_j)\mathbf{a}_j = \sum_j (\nabla f \times \mathbf{a}_j)\mathbf{a}_j = \nabla f \times \bar{\bar{\mathbf{I}}}, \tag{7.53}$$

which is a dyadic.

7-3 DYADIC INTEGRAL THEOREMS

There are several integral theorems in dyadic analysis which can be derived by changing the vector functions in the vector Green's theorems to dyadic functions.

First Vector–Dyadic Green's Theorem

To avoid a conflict of notation between a vector "**a**" and the unit vector $\mathbf{a}_i$, now adopted for the Cartesian system, the first vector Green's theorem stated by (12) of Appendix C will be written in the following form:

$$\iiint_V [(\nabla \times \mathbf{P}) \cdot (\nabla \times \mathbf{Q}) - \mathbf{P} \cdot \nabla \times \nabla \times \mathbf{Q}] dV$$

$$= \oiint_S \mathbf{n} \cdot (\mathbf{P} \times \nabla \times \mathbf{Q}) dS. \tag{7.54}$$

We have purposely placed the function $\mathbf{Q}$ in the posterior position in (7.54), a practice which is used to change a vector to a dyadic. Consider now three distinct $\mathbf{Q}_j$ with $j = 1, 2, 3$ so that we have three identities of the same form as (7.54). By juxtaposing a unit vector $\mathbf{a}_j$ at the posterior position of each of the three equations and summing them, we obtain

$$\iiint_V [(\nabla \times \mathbf{P}) \cdot (\nabla \times \bar{\bar{\mathrm{Q}}}) - \mathbf{P} \cdot \nabla \times \nabla \times \bar{\bar{\mathrm{Q}}}] dV$$

$$= \oiint_S \mathbf{n} \cdot (\mathbf{P} \times \nabla \times \bar{\bar{\mathrm{Q}}}) dS \tag{7.55}$$

where, by definition,

$$\nabla \times \bar{\bar{\mathrm{Q}}} = \sum_j (\nabla \times \mathbf{Q}_j) \mathbf{a}_j \tag{7.56}$$

and

$$\nabla \times \nabla \times \bar{\bar{\mathrm{Q}}} = \sum_j (\nabla \times \nabla \times \mathbf{Q}_j) \mathbf{a}_j. \tag{7.57}$$

Equation (7.55) is designated as the first vector–dyadic Green's theorem of Type A because it involves a vector function **P** and a dyadic function $\bar{\bar{\mathrm{Q}}}$. By interchanging **P** with **Q** in (7.54) and then raising the level of **Q** to a dyadic, we obtain

$$\iiint_V [(\nabla \times \mathbf{P}) \cdot (\nabla \times \bar{\bar{\mathrm{Q}}}) - (\nabla \times \nabla \times \mathbf{P}) \cdot \bar{\bar{\mathrm{Q}}}] dV$$

$$= - \oiint_S \mathbf{n} \cdot [(\nabla \times \mathbf{P}) \times \bar{\bar{\mathrm{Q}}}] dS. \tag{7.58}$$

Equation (7.58) is designated as the first vector–dyadic Green's theorem of Type B. Except for the term $(\nabla \times \mathbf{P}) \cdot (\nabla \times \bar{\bar{\mathrm{Q}}})$, which is common in (7.55) and (7.58), the rest are different.

Second Vector–Dyadic Green's Theorem

By subtracting (7.55) from (7.58), we obtain the second vector–dyadic Green's theorem:

$$\iiint_V [\mathbf{P} \cdot \nabla \times \nabla \times \bar{\bar{\mathrm{Q}}} - (\nabla \times \nabla \times \mathbf{P}) \cdot \bar{\bar{\mathrm{Q}}}] dV$$

$$= - \oiint_S \mathbf{n} \cdot [(\mathbf{P} \times \nabla \times \bar{\bar{\mathrm{Q}}}) + (\nabla \times \mathbf{P}) \times \bar{\bar{\mathrm{Q}}}] dS. \tag{7.59}$$

This theorem is probably the most useful formula in the application of dyadic analysis to electromagnetic theory [14].

First Dyadic–Dyadic Green's Theorem

Equation (7.55) can be elevated to a higher level by moving **P** and $\nabla \times \mathbf{P}$ into the posterior position and transposing the dyadic terms into the anterior position so that

$$\iiint_V \left\{[\nabla \times \bar{\bar{\mathrm{Q}}}]^T \cdot (\nabla \times \mathbf{P}) - [\nabla \times \nabla \times \bar{\bar{\mathrm{Q}}}]^T \cdot \mathbf{P}\right\} dV$$

$$= \oiint_S [\nabla \times \bar{\bar{\mathrm{Q}}})]^T \cdot (\mathbf{n} \times \mathbf{P}) dS. \tag{7.60}$$

The vector function **P** can now be elevated to a dyadic level that yields the first dyadic–dyadic Green's theorem:

$$\iiint_V \left\{[\nabla \times \bar{\bar{\mathrm{Q}}}]^T \cdot (\nabla \times \bar{\bar{\mathrm{P}}}) - [\nabla \times \nabla \times \bar{\bar{\mathrm{Q}}}]^T \cdot \bar{\bar{\mathrm{P}}}\right\} dV$$

$$= \oiint_S [\nabla \times \bar{\bar{\mathrm{Q}}})]^T \cdot (\mathbf{n} \times \bar{\bar{\mathrm{P}}}) dS. \tag{7.61}$$

By doing the same thing with (7.58), we obtain

$$\iiint_V \left\{[\nabla \times \bar{\bar{\mathrm{Q}}}]^T \cdot (\nabla \times \bar{\bar{\mathrm{P}}}) - [\bar{\bar{\mathrm{Q}}}]^T \cdot (\nabla \times \nabla \times \bar{\bar{\mathrm{P}}})\right\} dV$$

$$= - \oiint_S [\bar{\bar{\mathrm{Q}}}]^T \cdot (\mathbf{n} \times \nabla \times \bar{\bar{\mathrm{P}}}) dS. \tag{7.62}$$

Second Dyadic–Dyadic Green's Theorem

By taking the difference between (7.61) and (7.62), we obtain the second dyadic–dyadic Green's theorem:

$$\iiint_V \left\{[\nabla \times \nabla \times \bar{\bar{\mathrm{Q}}}]^T \cdot \bar{\bar{\mathrm{P}}} - [\bar{\bar{\mathrm{Q}}}]^T \cdot (\nabla \times \nabla \times \bar{\bar{\mathrm{P}}}\right\} dV$$

$$= - \oiint_S \left\{[\nabla \times \bar{\bar{\mathrm{Q}}})]^T \cdot (\mathbf{n} \times \bar{\bar{\mathrm{P}}}) + [\bar{\bar{\mathrm{Q}}}]^T \cdot (\mathbf{n} \times \nabla \times \bar{\bar{\mathrm{P}}})\right\} dS. \tag{7.63}$$

The two dyadic–dyadic Green's theorems involve two dyadics; hence, we have the name. They can be used to prove the symmetrical property of the electric and magnetic dyadic Green's functions [17]. In an earlier work [14], the proof was rather tedious by using the vector–dyadic Green's theorem because the dyadic–dyadic Green's theorems were not available then. We have now assembled all of the important formulas in dyadic analysis, with the hope that they will be useful in digesting technical articles involving dyadic analysis, particularly in its application to electromagnetic theory.

APPENDIX A

Transformation Between Unit Vectors

Cylindrical System

$$(v_1, v_2, v_3) = (r, \phi, z)$$
$$(h_1, h_2, h_3) = (1, r, 1)$$
$$x = r\,\cos\phi,\; y = r\,\sin\phi,\; z = z.$$

	$\mathbf{u}_x$	$\mathbf{u}_y$	$\mathbf{u}_z$
$\mathbf{u}_r$	$\cos\phi$	$\sin\phi$	0
$\mathbf{u}_\phi$	$-\sin\phi$	$\cos\phi$	0
$\mathbf{u}_z$	0	0	1

Spherical System

$$(v_1, v_2, v_3) = (R, \theta, \phi)$$
$$(h_1, h_2, h_3) = (1, R, R\,\sin\theta)$$
$$x = R\,\sin\theta\,\cos\phi,\; y = R\,\sin\theta\,\sin\phi,\; z = R\,\cos\theta.$$

	$\mathbf{u}_x$	$\mathbf{u}_y$	$\mathbf{u}_z$
$\mathbf{u}_R$	$\sin\theta\ \cos\phi$	$\sin\theta\ \sin\phi$	$\cos\theta$
$\mathbf{u}_\theta$	$\cos\theta\ \cos\phi$	$\cos\theta\ \sin\phi$	$-\sin\theta$
$\mathbf{u}_\phi$	$-\sin\phi$	$\cos\phi$	0

Elliptical Cylinder

$$(v_1, v_2, v_3) = (\eta, \xi, z)$$

$$(h_1, h_2, h_3) = \left[c\left(\frac{\xi^2 - \eta^2}{1 - \eta^2}\right)^{\frac{1}{2}},\ c\left(\frac{\xi^2 - \eta^2}{\xi^2 - 1}\right)^{\frac{1}{2}},\ 1\right]$$

$$x = c\,\eta\,\xi,\ y = c\left[\left(1 - \eta^2\right)\left(\xi^2 - 1\right)\right]^{\frac{1}{2}},\ z = z.$$

	$\mathbf{u}_x$	$\mathbf{u}_y$	$\mathbf{u}_z$
$\mathbf{u}_\eta$	$\frac{c\xi}{h_1}$	$\frac{-c\eta}{h_2}$	0
$\mathbf{u}_\xi$	$\frac{c\eta}{h_2}$	$\frac{c\xi}{h_1}$	0
$\mathbf{u}_z$	0	0	1

Parabolic Cylinder

$$(v_1, v_2, v_3) = (\eta, \xi, z)$$

$$(h_1, h_2, h_3) = \left[\left(\eta^2 + \xi^2\right)^{\frac{1}{2}},\ \left(\eta^2 + \xi^2\right)^{\frac{1}{2}},\ 1\right]$$

$$x = \frac{1}{2}\left(\eta^2 - \xi^2\right),\ y = \eta\xi,\ z = z.$$

	u_x	$\mathbf{u}_y$	$\mathbf{u}_z$
$\mathbf{u}_\eta$	$\frac{\eta}{h}$	$\frac{\xi}{h}$	0
$\mathbf{u}_\xi$	$\frac{-\xi}{h}$	$\frac{\eta}{h}$	0
$\mathbf{u}_z$	0	0	1

$$h = h_1 = h_2 = \left(\eta^2 + \xi^2\right)^{\frac{1}{2}}.$$

Prolate Spheroid

$$(v_1, v_2, v_3) = (\eta, \xi, \phi)$$

$$(h_1, h_2, h_3) = \left[c\left(\frac{\xi^2 - \eta^2}{1 - \eta^2}\right)^{\frac{1}{2}}, c\left(\frac{\xi^2 - \eta^2}{\xi^2 - 1}\right)^{\frac{1}{2}}, c\left(1 - \eta^2\right)^{\frac{1}{2}}\left(\xi^2 - 1\right)^{\frac{1}{2}}\right]$$

$$r = \left(x^2 + y^2\right)^{\frac{1}{2}} = c\left[\left(1 - \eta^2\right)\left(\xi^2 - 1\right)\right]^{\frac{1}{2}}, \ \phi = \phi$$

$$z = c\,\eta\,\xi.$$

	$\mathbf{u}_z$	$\mathbf{u}_r$	$\mathbf{u}_\phi$
$\mathbf{u}_\eta$	$\frac{c\xi}{h_1}$	$\frac{-c\eta}{h_2}$	0
$\mathbf{u}_\xi$	$\frac{c\eta}{h_2}$	$\frac{c\xi}{h_1}$	0
$\mathbf{u}_\phi$	0	0	1

The unit vectors $\mathbf{u}_r$ and $\mathbf{u}_\phi$ can be expressed in terms of $\mathbf{u}_x$ and $\mathbf{u}_y$ covered in the cylindrical case; the same applies to the oblate spheroid.

Oblate Spheroid

$$(v_1, v_2, v_3) = (\eta, \xi, \phi)$$

$$(h_1, h_2, h_3) = \left[c\left(\frac{\xi^2 - \eta^2}{\xi^2 - 1}\right)^{\frac{1}{2}}, c\left(\frac{\xi^2 - \eta^2}{1 - \eta^2}\right)^{\frac{1}{2}}, c\,\xi\,\eta\right]$$

$$r = \left(x^2 + y^2\right)^{\frac{1}{2}} = c\,\xi\,\eta$$

$$z = c\left[\left(\xi^2 - 1\right)\left(1 - \eta^2\right)\right]^{\frac{1}{2}}.$$

	$\mathbf{u}_z$	$\mathbf{u}_r$	$\mathbf{u}_\phi$
$\mathbf{u}_\xi$	$\frac{c\xi}{h_2}$	$\frac{c\eta}{h_1}$	0
$\mathbf{u}_\eta$	$\frac{-c\eta}{h_1}$	$\frac{c\xi}{h_2}$	0
$\mathbf{u}_\phi$	0	0	1

Bipolar Cylinders

$$(v_1, v_2, v_3) = (\eta, \xi, \phi)$$

$$(h_1, h_2, h_3) = \left(\frac{a}{\cosh\xi - \cosh\eta}, \frac{a}{\cosh\xi - \cos\eta}, 1\right)$$

$$x = \frac{a\ \sinh\xi}{\cosh\xi - \cos\eta},\ y = \frac{a\ \sin\eta}{\cosh\xi - \cos\eta},\ z = z.$$

	$\mathbf{u}_x$	$\mathbf{u}_y$	$\mathbf{u}_z$
$\mathbf{u}_\eta$	$\frac{-h}{a}\sinh\xi\ \sin\eta$	$\frac{h}{a}(\cosh\xi\ \cos\eta - 1)$	0
$\mathbf{u}_\xi$	$\frac{-h}{a}(\cosh\xi\ \cos\eta - 1)$	$\frac{-h}{a}\sinh\xi\ \sin\eta$	0
$\mathbf{u}_z$	0	0	1

where

$$h = h_1 = h_2 = \frac{a}{\cosh\xi - \cos\eta}.$$

In all of these tables, the unit vectors are all arranged in the order of a right-handed system, i.e., $\mathbf{u}_1 \times \mathbf{u}_2 = \mathbf{u}_3$.

APPENDIX B

Vector and Dyadic Identities

Vector Identities

$$\mathbf{a} \cdot (\mathbf{b} \times \mathbf{c}) = \mathbf{b} \cdot (\mathbf{c} \times \mathbf{a}) = \mathbf{c} \cdot (\mathbf{a} \times \mathbf{b}) \tag{B.1}$$

$$\mathbf{a} \times (\mathbf{b} \times \mathbf{c}) = (\mathbf{a} \cdot \mathbf{c})\,\mathbf{b} - (\mathbf{a} \cdot \mathbf{b})\,\mathbf{c} \tag{B.2}$$

$$\nabla (ab) = a\nabla b + b\nabla a \tag{B.3}$$

$$\nabla \cdot (a\mathbf{b}) = a\nabla \cdot \mathbf{b} + \mathbf{b} \cdot \nabla a \tag{B.4}$$

$$\nabla \times (a\mathbf{b}) = a\nabla \times \mathbf{b} - \mathbf{b} \times \nabla a \tag{B.5}$$

$$\nabla \cdot (\mathbf{a} \times \mathbf{b}) = \mathbf{b} \cdot \nabla \times \mathbf{a} - \mathbf{a} \cdot \nabla \times \mathbf{b} \tag{B.6}$$

$$\nabla (\mathbf{a} \cdot \mathbf{b}) = \mathbf{a} \times \nabla \times \mathbf{b} + \mathbf{b} \times \nabla \times \mathbf{a} + (\mathbf{a} \cdot \nabla)\,\mathbf{b} + (\mathbf{b} \cdot \nabla)\,\mathbf{a} \tag{B.7}$$

$$\nabla \times (\mathbf{a} \times \mathbf{b}) = \mathbf{a}\nabla \cdot \mathbf{b} - \mathbf{b}\nabla \cdot \mathbf{a} - (\mathbf{a} \cdot \nabla)\,\mathbf{b} + (\mathbf{b} \cdot \nabla)\,\mathbf{a} \tag{B.8}$$

$$\nabla \cdot (\nabla a) = \nabla^2 a \tag{B.9}$$

$$\nabla \times (\nabla \times \mathbf{a}) = \nabla (\nabla \cdot \mathbf{a}) - \nabla^2 \mathbf{a} \tag{B.10}$$

$$\nabla \times (\nabla a) = 0 \tag{B.11}$$

$$\nabla \cdot (\nabla \times \mathbf{a}) = 0 \tag{B.12}$$

Dyadic Identities

$$\mathbf{a}\cdot\left(\mathbf{b}\times\overline{\overline{c}}\right) = -\mathbf{b}\cdot\left(\mathbf{a}\times\overline{\overline{c}}\right) = \left(\mathbf{a}\times\mathbf{b}\right)\cdot\overline{\overline{c}} \tag{B.13}$$

$$\mathbf{a}\times\left(\mathbf{b}\times\overline{\overline{c}}\right) = \mathbf{b}\left(\mathbf{a}\cdot\overline{\overline{c}}\right) - \left(\mathbf{a}\cdot\mathbf{b}\right)\overline{\overline{c}} \tag{B.14}$$

$$\nabla\cdot\left(a\overline{\overline{b}}\right) = a\nabla\cdot\overline{\overline{b}} + \left(\nabla a\right)\cdot\overline{\overline{b}} \tag{B.15}$$

$$\nabla\times\left(a\overline{\overline{b}}\right) = a\nabla\times\overline{\overline{b}} + \left(\nabla a\right)\times\overline{\overline{b}} \tag{B.16}$$

$$\nabla\cdot\left(\mathbf{a}\times\overline{\overline{b}}\right) = \left(\nabla\times\mathbf{a}\right)\cdot\overline{\overline{b}} - \mathbf{a}\cdot\left(\nabla\times\overline{\overline{b}}\right) \tag{B.17}$$

$$\nabla\times\left(\nabla\times\overline{\overline{a}}\right) = \nabla\left(\nabla\cdot\overline{\overline{a}}\right) - \nabla^2\,\overline{\overline{a}} \tag{B.18}$$

$$\nabla\cdot\left(\nabla\times\overline{\overline{a}}\right) = 0 \tag{B.19}$$

$$\mathbf{a}\cdot\overline{\overline{b}} = \left[\overline{\overline{b}}\right]^T\cdot\mathbf{a} \tag{B.20}$$

$$\mathbf{a}\times\overline{\overline{b}} = -\left\{\left[\overline{\overline{b}}\right]^T\times\mathbf{a}\right\}^T \tag{B.21}$$

$$\left[\overline{\overline{c}}\right]^T\cdot\left(\mathbf{a}\times\overline{\overline{b}}\right) = -\left[\mathbf{a}\times\overline{\overline{c}}\right]^T\cdot\overline{\overline{b}} \tag{B.22}$$

APPENDIX C

Integral Theorems

In this appendix, we use $\mathbf{u}_\ell$, $\mathbf{u}_m$, and $\mathbf{u}_n$ to denote the tangential and normal unit vectors associated with a surface. $\mathbf{u}_n$ is normal to an open or a closed surface. $\mathbf{u}_\ell$ is tangential to the edge of an open surface. $\mathbf{u}_m$ is normal to the edge, but tangential to the surface. The triad forms an orthogonal set, i.e., $\mathbf{u}_m = \mathbf{u}_\ell \times \mathbf{u}_n$, and $d\mathbf{S} = \mathbf{u}_n \, dS, d\boldsymbol{\ell} = \mathbf{u}_\ell d\ell$.

1. Gauss or divergence theorem:

$$\iiint \nabla \cdot \mathbf{F} \, dV = \oiint (\mathbf{u}_n \cdot \mathbf{F}) \, dS.$$

2. Curl theorem:

$$\iiint \nabla \times \mathbf{F} \, dV = \oiint (\mathbf{u}_n \times \mathbf{F}) \, dS.$$

3. Gradient theorem:

$$\iiint \nabla f \, dV = \oiint \mathbf{u}_n f \, dS.$$

4. Surface divergence theorem:

$$\iint \nabla_s \cdot \mathbf{f} \, dS = \oint (\mathbf{u}_m \cdot \mathbf{f}) \, d\ell.$$

5. Surface curl theorem:

$$\iint \nabla_s \times \mathbf{f}\, dS = \oint (\mathbf{u}_m \times \mathbf{f})\, d\ell.$$

6. Surface gradient theorem:

$$\iint \nabla_s f\, dS = \oint \mathbf{u}_m f\, d\ell.$$

7. Cross-gradient theorem:

$$\iint \mathbf{u}_n \times \nabla f\, dS = \oint f d\boldsymbol{\ell}.$$

8. Stokes theorem:

$$\iint \mathbf{u}_n \cdot \nabla \times \mathbf{f}\, dS = \oint \mathbf{f} \cdot d\boldsymbol{\ell}.$$

9. Cross–del–cross theorem:

$$\iint (\mathbf{u}_n \times \nabla) \times \mathbf{f}\, dS = -\oint \mathbf{f} \times d\boldsymbol{\ell}.$$

10. First scalar Green's theorem:

$$\iiint [f_1\, \nabla^2 f_2 + (\nabla f_1) \cdot (\nabla f_2)]\, dV = \oiint f_1 \frac{\partial f_2}{\partial n} dS.$$

11. Second scalar Green's theorem:

$$\iiint (f_1\, \nabla^2 f_2 - f_2 \nabla^2 f_1)\, dV = \oiint \left(f_1 \frac{\partial f_2}{\partial n} - f_2 \frac{\partial f_1}{\partial n} \right) dS.$$

12. First vector Green's theorem:

$$\iiint [(\nabla \times \mathbf{F}_1) \cdot (\nabla \times \mathbf{F}_2) - \mathbf{F}_1 \cdot \nabla \times \nabla \times \mathbf{F}_2]\, dV$$
$$= \oiint \mathbf{u}_n \cdot (\mathbf{F}_1 \times \nabla \times \mathbf{F}_2)\, dS.$$

13. Second vector Green's theorem:

$$\iiint (\mathbf{F}_2 \cdot \nabla \times \nabla \times \mathbf{F}_1 - \mathbf{F}_1 \cdot \nabla \times \nabla \times \mathbf{F}_2)\, dV$$
$$= \oiint \mathbf{u}_n \cdot (\mathbf{F}_1 \times \nabla \times \mathbf{F}_2 - \mathbf{F}_2 \times \nabla \times \mathbf{F}_1)\, dS.$$

14. First vector–dyadic Green's theorem:

$$\iiint \left[(\nabla \times \mathbf{P}) \cdot \nabla \times \overline{\overline{Q}} - \mathbf{P} \cdot \nabla \times \nabla \times \overline{\overline{Q}} \right] dv$$

$$= \oiint \mathbf{u}_n \cdot \left(\mathbf{P} \times \nabla \times \overline{\overline{Q}} \right) dS.$$

15. Second vector–dyadic Green's theorem:

$$\iiint \left[(\nabla \times \nabla \times \mathbf{P}) \cdot \overline{\overline{Q}} - \mathbf{P} \cdot \nabla \times \nabla \times \overline{\overline{Q}} \right] dv$$

$$= \oiint \mathbf{u}_n \cdot \left[\mathbf{P} \times \left(\nabla \times \overline{\overline{Q}} \right) + (\nabla \times \mathbf{P}) \times \overline{\overline{Q}} \right] dS.$$

16. First dyadic–dyadic Green's theorem:

$$\iiint \left\{ \left[\overline{\overline{Q}} \right]^T \cdot \nabla \times \nabla \times \overline{\overline{P}} - \left[\nabla \times \overline{\overline{Q}} \right]^T \cdot \nabla \times \overline{\overline{P}} \right\} dV$$

$$= \oiint \left[\overline{\overline{Q}} \right]^T \cdot \left(\mathbf{u}_n \times \nabla \times \overline{\overline{P}} \right) dS.$$

17. Second dyadic–dyadic Green's theorem:

$$\iiint \left\{ \left[\overline{\overline{Q}} \right]^T \cdot \nabla \times \nabla \times \overline{\overline{P}} - \left[\nabla \times \nabla \times \overline{\overline{Q}} \right]^T \cdot \overline{\overline{P}} \right\} dV$$

$$= \oiint \left\{ \left[\overline{\overline{Q}} \right]^T \cdot \left(\mathbf{u}_n \times \nabla \times \overline{\overline{P}} \right) + \left[\nabla \times \overline{\overline{Q}} \right]^T \cdot \left(\mathbf{u}_n \times \overline{\overline{P}} \right) \right\} dS.$$

18. Helmholtz transport theorem:

$$\frac{d}{dt} \iint_{S(t)} \mathbf{F} \cdot d\mathbf{S} = \iint_{S(t)} \left[\frac{\partial \mathbf{F}}{\partial t} + \mathbf{v} \nabla \cdot \mathbf{F} - \nabla \times (\mathbf{v} \times \mathbf{F}) \right] dS.$$

19. Maxwell theorem:

$$\frac{d}{dt} \oint_{L(t)} \mathbf{f} \cdot d\boldsymbol{\ell} = \oint_{L(t)} \left(\frac{\partial \mathbf{f}}{\partial t} - \mathbf{v} \times \nabla \times \mathbf{f} \right) \cdot d\boldsymbol{\ell}.$$

20. Reynolds transport theorem:

$$\frac{d}{dt} \iiint_{V(t)} \rho dV = \iiint_{V(t)} \left[\frac{\partial \rho}{\partial t} + \nabla \cdot (\rho \mathbf{v}) \right] dV.$$

APPENDIX D

Relationships Between Integral Theorems

The integral theorems stated by (1–3) and (7–9) in Appendix C are closely related. By means of the gradient theorem, we can derive both the divergence theorem and the curl theorem. From this point of view, we must first prove the gradient theorem, leaving aside its derivation by the symbolic method. The theorem states that

$$\iiint \nabla f \, dV = \oint\!\!\!\oint_S f \, d\mathbf{S}. \tag{D.1}$$

In a Cartesian system with coordinate variables (x_1, x_2, x_3), the x_1 component of (D.1) corresponds to

$$\iiint \frac{\partial f}{\partial x_1} dx_1 dx_2 dx_3 = \iint_{S_2} f dx_2 dx_3 - \iint_{S_1} f dx_2 dx_3 \tag{D.2}$$

where S_1 and S_2 denote the two sides of an enclosed surface S viewed in the x_1 direction. The negative sign associated with the surface integral evaluated on S_1 is due to the fact that the vector component of $d\mathbf{S}_1$ is equal to $-dx_2 dx_3 \mathbf{a}_1$. Equation (D.2) is a valid identity because the volume integral is given by

$$\begin{aligned} \iiint \frac{\partial f}{\partial x_1} dx_1 dx_2 dx_3 &= \iint \left[f\left(P_2\right) - f\left(P_1\right) \right] dx_2 dx_3 \\ &= \iint_{S_2} f dx_2 dx_3 - \iint_{S_1} f dx_2 dx_3 \end{aligned} \tag{D.3}$$

where P_2 and P_1 denote two stations located at opposite sides of the surface along the x_1 direction. The same procedure can be used to prove the remaining two components of (D.1). Having proved the validity of (D.1), we can use it to deduce the divergence theorem (Gauss theorem) and the curl theorem.

We now consider three distinct sets of (D.1) in the form

$$\iiint \nabla F_i dV = \oiint_S F_i d\mathbf{S}, \qquad i = 1, 2, 3. \tag{D.4}$$

By taking the scalar product of (D.4) with $\mathbf{a}_i$ and summing the resultant equations, we obtain

$$\sum_i \mathbf{a}_i \cdot \iiint \nabla F_i dV = \sum_i \mathbf{a}_i \cdot \oiint_S F_i d\mathbf{S}. \tag{D.5}$$

Let

$$\mathbf{F} = \sum F_i \mathbf{a}_i,$$

and since

$$\mathbf{a}_i \cdot \nabla F_i = \nabla \cdot (F_i \mathbf{a}_i),$$

we obtain

$$\iiint \nabla \cdot \mathbf{F} dV = \oiint_S \mathbf{F} \cdot d\mathbf{S}, \tag{D.6}$$

which is the divergence theorem. Similarly, by taking the cross product of $\mathbf{a}_i$ with (D.4), we obtain

$$\sum_i \mathbf{a}_i \times \iiint \nabla F_i dV = \sum_i \mathbf{a}_i \times \oiint F_i d\mathbf{S}. \tag{D.7}$$

Since

$$\mathbf{a}_i \times \nabla F_i = -\nabla \times (F_i \mathbf{a}_i),$$

(D.7) is equivalent to

$$\iiint \nabla \times \mathbf{F} dV = -\oiint \mathbf{F} \times d\mathbf{S}, \tag{D.8}$$

which is the curl theorem. The approach which we took can be applied to the other three theorems listed as 7–9 in Appendix C. In this case, we consider the cross-gradient theorem as the key theorem which must be proved first. The theorem states that

$$\iint \mathbf{u}_n \times \nabla f dS = \oint f d\boldsymbol{\ell}. \tag{D.9}$$

In a Cartesian system, we can write

$$\mathbf{u}_n = \sum_i n_i \mathbf{a}_i$$

$$d\boldsymbol{\ell} = \sum_i dx_i \mathbf{a}_i.$$

Then the x_1 component of (D.9) reads

$$\iint \left(n_2 \frac{\partial f}{\partial x_3} - n_3 \frac{\partial f}{\partial x_2} \right) dS = \oint f dx_1. \tag{D.10}$$

The surface integral in (D.10) can be written in the form

$$\begin{aligned}
&\iint \left(\frac{\partial f}{\partial x_3} dx_1 dx_3 + \frac{\partial f}{\partial x_2} dx_1 dx_2 \right) \\
&= \iint \left(\frac{\partial f}{\partial x_2} dx_2 + \frac{\partial f}{\partial x_3} dx_3 \right) dx_1 \\
&= \iint df dx_1 = \int \left[f\left(p_2\right) - f\left(p_1\right) \right] dx_1 \\
&= \int_{c_2} f\left(p_2\right) dx_1 - \int_{-c_1} f\left(p_1\right) dx_1 \\
&= \oint_c f dx_1
\end{aligned}$$

where p_1 and p_2 denote two stations on the closed contour, which consists of two segments $c_1 + c_2$. We have thus proved the validity of (D.10). The same procedure applies to the x_2 and x_3 components of (D.9). Once we have proved the cross-gradient theorem, it can be used to deduce the Stokes theorem and the cross–del–cross theorem.

We consider three distinct sets of (D.9) in the form

$$\iint \mathbf{u}_n \times \nabla F_i dS = \oint F_i d\boldsymbol{\ell}, \qquad i = 1, 2, 3. \tag{D.11}$$

By taking the scalar product of (D.11) with $\mathbf{a}_i$ and summing the resultant equations, we obtain

$$\sum_i \mathbf{a}_i \cdot \iint \left(\mathbf{u}_n \times \nabla F_i \right) dS = \sum_i \mathbf{a}_i \cdot \oint F_i d\boldsymbol{\ell}. \tag{D.12}$$

Since

$$\begin{aligned}
\mathbf{a}_i \cdot \left(\mathbf{u}_n \times \nabla F_i \right) &= -\mathbf{u}_n \cdot \left(\mathbf{a}_i \times \nabla F_i \right) \\
&= \mathbf{u}_n \cdot \nabla \times \left(F_i \mathbf{a}_i \right)
\end{aligned}$$

and we let

$$\mathbf{F} = \sum_i F_i \mathbf{a}_i,$$

(D.12) can be written in the form

$$\iint \mathbf{u}_n \cdot \nabla \times \mathbf{F} dS = \oint \mathbf{F} \cdot d\boldsymbol{\ell}, \tag{D.13}$$

which is the Stokes theorem. Similarly, by taking the cross product of (D.11) with $\mathbf{a}_i$, we obtain

$$\sum_i \mathbf{a}_i \times \iint (\mathbf{u}_n \times \nabla F_i)\, dS = \sum_i \mathbf{a}_i \times \oint F_i d\boldsymbol{\ell}$$

or

$$-\sum_i \iint (\mathbf{u}_n \times \nabla)\, F_i \times \mathbf{a}_i dS = \sum_i \oint F_i \mathbf{a}_i \times d\boldsymbol{\ell}. \qquad \text{(D.14)}$$

Hence,

$$\iint (\mathbf{u}_n \times \nabla) \times \mathbf{F} dS = -\oint \mathbf{F} \times d\boldsymbol{\ell}, \qquad \text{(D.15)}$$

which corresponds to the cross–del–cross theorem. It is seen that in this analysis, the gradient theorem and the cross-gradient theorem are considered the key theorems based on which the other four theorems can be readily derived. The approach taken here has its own merit without considering the derivation of these theorems, independently, by the symbolic method.

The relationships among the gradient theorem, the divergence theorem, and the curl theorem have previously been pointed out by Van Bladel [1, Appendix I]. Alternatively, we can use the divergence theorem and the Stokes theorem as the key theorems to derive the other four theorems. The manipulations, however, are more involved.

APPENDIX E

Differential Operators in Vector Analysis and the Laplacian of a Vector in the Curvilinear Orthogonal System

Abstract—Some long-existing misunderstandings of the meaning of the del operator in the curvilinear orthogonal system have been pointed out in this work. One misunderstanding results from a false manipulation of the notations for the divergence and the curl introduced by Gibbs. A proper analysis shows that there are three distinct differential operators in a curvilinear orthogonal system, and the Laplacian of a vector function is a well-defined entity.

E-1 INTRODUCTION

Vector analysis is an indispensable tool in the teaching of electromagnetics, hydrodynamics, and mechanics. Unfortunately, there have been some misunderstandings which have been in existence for a long time. In this work, we attempt to clarify them by a critical examination of these problems.

The del operator (or the Nabla operator or Hamilton operator) in a Cartesian system is defined by

$$\nabla = \sum_i \mathbf{a}_i \frac{\partial}{\partial x_i} \tag{E.1}$$

where x_i and $\mathbf{a}_i$ with $i = (1, 2, 3)$ denote, respectively, the coordinate variables and the unit vectors in that system. The gradient of a scalar function f can then be written as

This appendix is a reproduction of Technical Report RL859 bearing the same title issued by the Radiation Laboratory, University of Michigan, Ann Arbor, MI, December 1990.

$$\nabla f = \sum_i \mathbf{a}_i \frac{\partial f}{\partial x_i} \tag{E.2}$$

which has no ambiguity, provided we accept the distributive rule in such an operation. For the divergence and the curl of a vector function **f**, Gibbs, one of the pioneers in vector analysis, introduced the <u>notations</u> $\nabla \cdot \mathbf{f}$ and $\nabla \times \mathbf{f}$ for these two functions, and <u>defined</u> them as [E3]

$$\nabla \cdot \mathbf{f} = \sum \mathbf{a}_i \cdot \frac{\partial \mathbf{f}}{\partial x_i} = \sum_i \frac{\partial f_i}{\partial x_i} \tag{E.3}$$

$$\nabla \times \mathbf{f} = \sum_i \mathbf{a}_i \times \frac{\partial \mathbf{f}}{\partial x_i} = \sum_i \left(\frac{\partial f_k}{\partial x_j} - \frac{\partial f_j}{\partial x_k} \right) \mathbf{a}_i \tag{E.4}$$

where $(i, j, k) = (1, 2, 3)$ in cyclic order. These are well-known expressions.

It should be pointed out that if one treats $\nabla \cdot \mathbf{f}$, the notation for the divergence, as the "scalar product" between ∇ and **f**, then

$$\nabla \cdot \mathbf{f} = \left(\sum_i \mathbf{a}_i \frac{\partial}{\partial x_i} \right) \cdot \mathbf{f}. \tag{E.5}$$

This is meaningless because the member on the right side of (E.5) consists of an assembly of functions and symbols, and is not a mathematically meaningful expression. We cannot arbitrarily transport the dot to the front of the differential sign nor transport the vector $\mathbf{a}_i$ behind the differential sign in order to create a meaningful expression of our choice. This is not a matter of interpretation; it is a false manipulation. We are not allowed to do this in mathematics. The situation is as if we have an assembly of numbers and signs in the form of $2 + \times 3$, which has no meaning in arithmetic. But if we move the plus sign to the front, we create a well-defined number $+2 \times 3$, and if we move the plus sign to the back, we create a numerical operator $(2 \times 3) + = 6+$. Neither of these expressions is equivalent to the original assembly. Unfortunately, many authors treat (E.5) to be equivalent to (E.3), and this creates a lot of confusion and many misunderstandings. For example, Moon and Spencer [E5, Appendix C, p. 324] state: "... A scalar product (between ∇ and **f**) gives the divergence

$$\operatorname{div} \mathbf{f} = \nabla \cdot \mathbf{f} = \frac{\partial f_x}{\partial x} + \frac{\partial f_y}{\partial y} + \frac{\partial f_z}{\partial z} \text{ ."}$$

We have changed their notation for the vector function to **f**. They did the same for $\nabla \times \mathbf{f}$, treating it as a "vector product" between ∇ and **f**. Later, they apply the same "interpretation" to $\nabla \cdot \mathbf{f}$ in a curvilinear orthogonal system that leads them to a wrong conclusion. We should emphasize here that Gibbs introduced $\nabla \cdot \mathbf{f}$ and $\nabla \times \mathbf{f}$ merely as the <u>notations</u> for the divergence and the curl, and they were not meant to be the "scalar product" and "vector product" between ∇ and **f**. The false manipulation was imposed upon these notations by later workers. There are dozens of authors who did the same as Moon and Spencer. Most of them are authors of books on electromagnetics, vector analysis, and calculus. The history behind this misunderstanding is long and interesting. The story is covered in [E10].

In a curvilinear orthogonal system with the coordinate variable denoted by v_i, the unit vector by $\mathbf{u}_i$, and the metric coefficients by h_i, the del operator is defined by

$$\nabla = \sum_i \frac{\mathbf{u}_i}{h_i} \frac{\partial}{\partial v_i}. \tag{E.6}$$

The gradient, still denoted by ∇f, can be written in the form

$$\nabla f = \sum_i \frac{\mathbf{u}_i}{h_i} \frac{\partial f}{\partial v_i}. \tag{E.7}$$

It is understood that the distributive rule has been forced upon the operand of the del operator. The meaning of ∇ in (E.7) has no ambiguity. When this vector differential operator is applied to a scalar function, the result yields the gradient of that function. It is the use of the del operator in divergence and curl that has created many problems.

For example, in the book by Morse and Feshbach [E7, part 1, p. 44], we find the following statement: "...The vector operator must have different forms for its different uses:

$$\begin{aligned} \nabla &= \sum_i \frac{\mathbf{u}_i}{h_i} \frac{\partial}{\partial v_i} && \text{for the gradient} \\ &= \frac{1}{\Omega} \sum_i \mathbf{u}_i \frac{\partial}{\partial v_i} \left(\frac{\Omega}{h_i} \right) && \text{for the divergence} \end{aligned}$$

and no form which can be written for the curl."

We have used Ω to represent $h_1 h_2 h_3$, and have changed their coordinate variables ξ_i to v_i and their notations $\mathbf{a}_i$ to $\mathbf{u}_i$. It is obvious that the "operator" introduced by these two authors for the divergence can produce the correct differential expression for the divergence only if the operation is "interpreted" as

$$\left[\frac{1}{\Omega} \sum_i \mathbf{u}_i \frac{\partial}{\partial v_i} \left(\frac{\Omega}{h_i} \right) \right] \cdot \mathbf{f} = \frac{1}{\Omega} \sum_i \frac{\partial}{\partial v_i} \left(\frac{\Omega}{h_i} \mathbf{u}_i \cdot \mathbf{f} \right). \tag{E.8}$$

Such an interpretation is quite arbitrary, and it does not follow the accepted rule of a differential operator because their ∇ for the divergence is a differentiated function, not an operator.

In the book by Moon and Spencer [E5, pp. 325–326], the two authors, presumably following their notion that $\nabla \cdot \mathbf{f}$ is the "scalar product" between ∇ and $\mathbf{f}$, interpreted $\nabla \cdot \mathbf{f}$ as

$$\nabla \cdot \mathbf{f} = \sum_i \frac{1}{h_i} \frac{\partial}{\partial u_i} (\mathbf{u}_i \cdot \mathbf{f}) \tag{E.9}$$

and then concluded that it does not yield the correct result. Furthermore, in commenting on the work by Phillips [E8], they asserted that Phillips manages to use the del operator in a curvilinear system for divergence and curl, but only by the trick of redefining the scalar and vector products for these particular applications. Actually, Phillips' method is not a trick at all; it is very ingenious, and he did not redefine the scalar and vector products under consideration. He obtains the correct expressions for the divergence and curl based on the differential expression for the gradient and some vector identities. By doing so, there is no need for him to discuss the role played by the del operator in $\nabla \cdot \mathbf{f}$ and $\nabla \times \mathbf{f}$ when they are expressed in the curvilinear system. The preceding introduction clearly indicates that we need a better understanding of the role played by the del operator in a curvilinear system.

E-2 THE DIFFERENTIAL OPERATORS

To avoid repetition in writing equations of a similar form, we will introduce a unified definition of the three key functions in one formula [E2], [E1, Appendix II, p. 477], [E9] which is independent of the coordinate system. The formula can be written in the form

$$\nabla * \tilde{f} = \lim_{\Delta V \to 0} \frac{\sum_j \left(\mathbf{n}_j * \tilde{f}\right) \Delta S_j}{\Delta V}. \tag{E.10}$$

The meaning of the asterisk "$*$" and the function $\tilde{f}$ with the tilde is as shown in Table E-1.

TABLE E-1

$*$	$\tilde{f}$	$\nabla * \tilde{f}$	Name
Null	f	∇f	Gradient
$\cdot$	$\mathbf{f}$	$\nabla \cdot \mathbf{f}$	Divergence
$\times$	$\mathbf{f}$	$\nabla \times \mathbf{f}$	Curl

In (E.10), $\mathbf{n}_j$ denotes a typical unit vector pointed outward from a surface $\Delta \mathbf{S}_j = \mathbf{n}_j \Delta S_j$, which is a part of the surface enclosing an elementary volume ΔV. By considering an elementary volume bounded by the constant coordinate surfaces in a curvilinear orthogonal system and taking the limit of (E.10), we obtain

$$\begin{aligned} \nabla * \tilde{f} &= \frac{1}{\Omega} \sum_i \frac{\partial}{\partial v_i} \left(\frac{\Omega}{h_i} \mathbf{u}_i * \tilde{f} \right) \\ &= \frac{1}{\Omega} \sum_i \left[\frac{\partial}{\partial v_i} \left(\frac{\Omega}{h_i} \mathbf{u}_i \right) * \tilde{f} + \frac{\Omega}{h_i} \mathbf{u}_i * \frac{\partial \tilde{f}}{\partial v_i} \right] \end{aligned} \tag{E.11}$$

where $\Omega = h_1 h_2 h_3$, as before.

It is known that the derivatives of the unit vectors in a curvilinear orthogonal system satisfy the following relationships [E7, pp. 25–26]:

$$\frac{\partial \mathbf{u}_j}{\partial v_i} = \frac{\partial h_i}{h_j \partial v_j}\mathbf{u}_i, \qquad i \neq j \tag{E.12}$$

$$\frac{\partial \mathbf{u}_i}{\partial v_i} = -\left(\frac{\partial h_i}{h_j \partial v_j}\mathbf{u}_j + \frac{\partial h_i}{h_k \partial v_k}\mathbf{u}_k\right) \tag{E.13}$$

with $(i, j, k) = (1, 2, 3)$ in cyclic order. Based on these relationships, it can be shown that

$$\sum_i \frac{\partial}{\partial v_i}\left(\frac{\Omega}{h_i}\mathbf{u}_i\right) = 0. \tag{E.14}$$

Equation (E.11) then reduces to

$$\nabla * \tilde{f} = \sum_i \frac{\mathbf{u}_i}{h_i} * \frac{\partial \tilde{f}}{\partial v_i}. \tag{E.15}$$

The differential form of the three key functions, therefore, can be written as

$$\nabla f = \sum_i \frac{\mathbf{u}_i}{h_i}\frac{\partial f}{\partial v_i} \tag{E.16}$$

$$\nabla \cdot \mathbf{f} = \sum_i \frac{\mathbf{u}_i}{h_i} \cdot \frac{\partial \mathbf{f}}{\partial v_i} \tag{E.17}$$

$$\nabla \times \mathbf{f} = \sum_i \frac{\mathbf{u}_i}{h_i} \times \frac{\partial \mathbf{f}}{\partial v_i}. \tag{E.18}$$

When applying these formulas to a Cartesian system with $h_i = 1, v_i = x_i, \mathbf{u}_i = \mathbf{a}_i$, we obtain the expressions used by Gibbs to <u>define</u> these functions. The proper meaning of the operators in the curvilinear system is now shown explicitly in (E.16)–(E.18). There are three distinct differential operators involved. For the gradient, we have the ordinary del operator. For the divergence, we have a differential operator, which will be denoted by $\nabla\!\!\!\!\cdot\;$ and designated as the divergence operator or the dot–del operator. It is defined by

$$\nabla\!\!\!\!\cdot\; = \sum_i \frac{\mathbf{u}_i}{h_i} \cdot \frac{\partial}{\partial v_i}. \tag{E.19}$$

For the curl, we have another differential operator, which will be denoted by $\nabla\!\!\!\!\!\times\;$ and designated as the curl operator or the cross–del operator. It is defined by

$$\nabla\!\!\!\!\!\times\; = \sum_i \frac{\mathbf{u}_i}{h_i} \times \frac{\partial}{\partial v_i}. \tag{E.20}$$

The three key functions in vector analysis can now be written as ∇f, $\nabla\!\!\!\!\cdot\;\mathbf{f}$, and $\nabla\!\!\!\!\!\times\;\mathbf{f}$. These notations are very descriptive. There is, of course, very little hope that we can change the long-established notation of Gibbs. We shall still use Gibbs' notations in this work.

The derivatives of the vector function **f** in (E.17) and (E.18) can be evaluated explicitly to obtain the well-known differential expressions for these two functions. Thus,

$$\begin{aligned}\frac{\partial \mathbf{f}}{\partial v_i} &= \frac{\partial}{\partial v_i}\sum_j f_j \mathbf{u}_j \\ &= \sum_j \left(\frac{\partial f_i}{\partial v_i}\mathbf{u}_j + f_j \frac{\partial \mathbf{u}_j}{\partial v_i}\right).\end{aligned} \tag{E.21}$$

With the aid of (E.12) and (E.13), the derivatives of the unit vectors in (E.21) can be expressed in terms of the unit vectors themselves. After some straightforward reductions, we obtain

$$\nabla \cdot \mathbf{f} = \frac{1}{\Omega}\sum \frac{\partial}{\partial v_i}\left(\frac{\Omega}{h_i} f_i\right) \tag{E.22}$$

and

$$\nabla \times \mathbf{f} = \frac{1}{\Omega}\sum \left[\frac{\partial\left(h_k f_k\right)}{\partial v_j} - \frac{\partial\left(h_j f_j\right)}{\partial v_k}\right] h_i \mathbf{u}_i \tag{E.23}$$

where

$$(i, j, k) = (1, 2, 3) \text{ in cyclic order.}$$

The expression for both $\nabla \cdot \mathbf{f}$ and $\nabla \times \mathbf{f}$ can be obtained more conveniently from the second term of (E.11) before its decomposition [E1, Appendix II, p. 477], [E9]. With this much discussion of the meaning of the operators in a curvilinear system, we turn to another related subject dealing with the Laplacian of a vector function and its identity.

E-3 THE LAPLACIAN OF A VECTOR FUNCTION

In a Cartesian system, it is well known that the following identity exists:

$$\nabla \cdot \nabla \mathbf{F} = \nabla\left(\nabla \cdot \mathbf{F}\right) - \nabla \times \left(\nabla \times \mathbf{F}\right) \tag{E.24}$$

where

$$\nabla \cdot \nabla \mathbf{F} = \sum_i \left(\frac{\partial^2 F_i}{\partial x_i^2}\right)\mathbf{a}_i = \sum_i \left(\nabla^2 F_i\right)\mathbf{a}_i. \tag{E.25}$$

The Laplacian of the scalar functions F_i has the meaning of

$$\nabla^2 F_i = \nabla \cdot \left(\nabla F_i\right) = \text{ div grad } F_i. \tag{E.26}$$

The fact that (E.24) is an identity independent of the choice of the coordinate system, including the oblique system, has been proved by Ignatowsky [E4,

p. 29] and Javid and Brown [E1, Appendix IV, pp. 483–484], based on the integral representation of the del operator [E4, p. 16], [E2], corresponding to our (E.10). Ignatowsky's proof requires some additional interpretation for the curvilinear system, while the proof by Javid and Brown is clear-cut.

In this work, we will prove (E.24) more specifically in a curvilinear orthogonal system based on a functional analysis.

Let us first examine the structure of $\nabla \cdot \nabla \mathbf{F}$ based on the differential forms of the gradient, the divergence, and the curl, as described by (E.16)–(E.18). Thus,

$$\nabla \cdot \nabla \mathbf{F} = \sum_i \frac{\mathbf{u}_i}{h_i} \cdot \frac{\partial}{\partial v_i} \left(\sum_j \frac{\mathbf{u}_j}{h_j} \frac{\partial \mathbf{F}}{\partial v_j} \right). \tag{E.27}$$

It should be observed that $\nabla \mathbf{F}$ is not a vector; it is a dyadic defined by

$$\nabla \mathbf{F} = \sum_j \frac{\mathbf{u}_j}{h_j} \frac{\partial \mathbf{F}}{\partial v_j}. \tag{E.28}$$

The positioning of the two terms in (E.28) must be kept in that order. Then

$$\frac{\partial}{\partial v_i} \nabla \mathbf{F} = \sum_j \left[\frac{\mathbf{u}_j}{h_j} \frac{\partial^2 \mathbf{F}}{\partial v_i \partial v_j} + \frac{\partial}{\partial v_i} \left(\frac{\mathbf{u}_j}{h_j} \right) \frac{\partial \mathbf{F}}{\partial v_j} \right]; \tag{E.29}$$

hence,

$$\begin{aligned} \nabla \cdot \nabla \mathbf{F} &= \sum_i \sum_j \frac{\mathbf{u}_i}{h_i} \cdot \left[\frac{\mathbf{u}_j}{h_j} \frac{\partial^2 \mathbf{F}}{\partial v_i \partial v_j} + \frac{\partial}{\partial v_i} \left(\frac{\mathbf{u}_j}{h_j} \right) \frac{\partial \mathbf{F}}{\partial v_j} \right] \\ &= \sum_i \sum_j \left\{ \left(\frac{\mathbf{u}_i}{h_i} \cdot \frac{\mathbf{u}_j}{h_j} \right) \frac{\partial^2 \mathbf{F}}{\partial v_i \partial v_j} + \left[\frac{\mathbf{u}_i}{h_i} \cdot \frac{\partial}{\partial v_i} \left(\frac{\mathbf{u}_j}{h_j} \right) \right] \frac{\partial \mathbf{F}}{\partial v_j} \right\} \\ &= \sum_i \frac{1}{h_i^2} \frac{\partial^2 \mathbf{F}}{\partial v_i^2} + \sum_i \sum_j \left[\frac{\mathbf{u}_i}{h_i} \cdot \frac{\partial}{\partial v_i} \left(\frac{\mathbf{u}_j}{h_j} \right) \right] \frac{\partial \mathbf{F}}{\partial v_j}. \end{aligned} \tag{E.30}$$

By interchanging the roles of i and j in the last term of (E.30), and with the aid of (E.12) and (E.13), we have

$$\sum_i \sum_j \left[\frac{\mathbf{u}_j}{h_j} \cdot \frac{\partial}{\partial v_j} \left(\frac{\mathbf{u}_i}{h_i} \right) \right] = \sum_i \frac{1}{\Omega} \frac{\partial}{\partial v_i} \left(\frac{\Omega}{h_i^2} \right), \qquad \Omega = h_1\, h_2\, h_3.$$

Hence,

$$\begin{aligned} \nabla \cdot \nabla \mathbf{F} &= \sum_i \left[\frac{1}{h_i^2} \frac{\partial^2 \mathbf{F}}{\partial v_i^2} + \frac{1}{\Omega} \frac{\partial}{\partial v_i} \left(\frac{\Omega}{h_i^2} \right) \frac{\partial \mathbf{F}}{\partial v_i} \right] \\ &= \sum_i \frac{1}{\Omega} \frac{\partial}{\partial v_i} \left(\frac{\Omega}{h_i^2} \frac{\partial \mathbf{F}}{\partial v_i} \right), \end{aligned} \tag{E.31}$$

which is the Laplacian of $\mathbf{F}$, and which can be written in the form of $\nabla^2\mathbf{F}$, with the Laplacian operator defined by

$$\nabla^2 = \sum_i \frac{1}{\Omega}\frac{\partial}{\partial v_i}\left(\frac{\Omega}{h_i^2}\frac{\partial}{\partial v_i}\right). \tag{E.32}$$

The operator applies to the entire function of $\mathbf{F}$, including its scalar components and the unit vectors. Actually, this form can be obtained in a very simple way by starting with the differential form of $\nabla^2\mathbf{F}$ in a Cartesian system and then converting the Laplacian operator to its form in a curvilinear system as follows:

$$\begin{aligned}\nabla^2\mathbf{F} &= \sum_j \left(\nabla^2 F_{x_j}\right)\mathbf{a}_j = \sum_j\sum_i \frac{1}{\Omega}\frac{\partial}{\partial v_i}\left(\frac{\Omega}{h_i^2}\frac{\partial F_{x_j}}{\partial v_i}\right)\mathbf{a}_j \\ &= \sum \frac{1}{\Omega}\frac{\partial}{\partial v_i}\left(\frac{\Omega}{h_i^2}\frac{\partial \mathbf{F}}{\partial v_i}\right)\end{aligned} \tag{E.33}$$

where

$$\mathbf{F} = \sum_j F_{x_j}\mathbf{a}_j = \sum_j F_j\mathbf{u}_j.$$

The structure of $\nabla\cdot\nabla\mathbf{F}$ in the form of (E.30), however, is needed later to prove identity (E.24) in a curvilinear system.

The two functions on the right side of (E.24) can be written in the form

$$\begin{aligned}\nabla(\nabla\cdot\mathbf{F}) &= \sum_i \frac{\mathbf{u}_i}{h_i}\frac{\partial}{\partial v_i}\sum_j \frac{\mathbf{u}_j}{h_j}\cdot\frac{\partial \mathbf{F}}{\partial v_j} \\ &= \sum_i\sum_j \frac{\mathbf{u}_i}{h_i}\left[\frac{\partial}{\partial v_i}\left(\frac{\mathbf{u}_j}{h_j}\right)\cdot\frac{\partial \mathbf{F}}{\partial v_j} + \frac{\mathbf{u}_j}{h_j}\cdot\frac{\partial^2 \mathbf{F}}{\partial v_i \partial v_j}\right]\end{aligned} \tag{E.34}$$

$$\begin{aligned}\nabla\times(\nabla\times\mathbf{F}) &= \sum_i \frac{\mathbf{u}_i}{h_i}\times\frac{\partial}{\partial v_i}\left(\sum_j \frac{\mathbf{u}_j}{h_j}\times\frac{\partial \mathbf{F}}{\partial v_j}\right) \\ &= \sum_i\sum_j \frac{\mathbf{u}_i}{h_i}\times\left[\frac{\partial}{\partial v_i}\left(\frac{\mathbf{u}_j}{h_j}\right)\times\frac{\partial \mathbf{F}}{\partial v_j} + \frac{\mathbf{u}_j}{h_j}\times\frac{\partial^2 \mathbf{F}}{\partial v_i \partial v_j}\right].\end{aligned} \tag{E.35}$$

The triple products in (E.35) can be decomposed into vectors using the identity

$$\mathbf{c}\times(\mathbf{a}\times\mathbf{b}) = (\mathbf{c}\cdot\mathbf{b})\,\mathbf{a} - (\mathbf{c}\cdot\mathbf{a})\,\mathbf{b}. \tag{E.36}$$

The subtraction of (E.35) from (E.34) yields

$$\nabla(\nabla\cdot\mathbf{F}) - \nabla\times(\nabla\times\mathbf{F}) = \nabla\cdot\nabla\mathbf{F} + \sum_i\sum_j \frac{\partial \mathbf{F}}{\partial v_j}\times\left[\frac{\mathbf{u}_i}{h_i}\times\frac{\partial}{\partial v_i}\left(\frac{\mathbf{u}_j}{h_j}\right)\right] \tag{E.37}$$

where we have used (E.30) to represent two of the terms in that resultant equation, and the last term in (E.37) results from a recombination of another two terms. With the aid of (E.12) and (E.13), it can be shown that

$$\sum_i \left[\frac{\mathbf{u}_i}{h_i} \times \frac{\partial}{\partial v_i} \left(\frac{\mathbf{u}_j}{h_j} \right) \right] = 0. \tag{E.38}$$

Alternatively, we can treat (E.38) as a vector identity, viz.

$$\sum_i \frac{\mathbf{u}_i}{h_i} \times \frac{\partial}{\partial v_i} \left(\frac{\mathbf{u}_j}{h_j} \right) = \nabla \times \left(\frac{\mathbf{u}_j}{h_j} \right) = \nabla \times \nabla v_j = 0. \tag{E.39}$$

Hence, (E.37) reduces to the identity

$$\nabla \cdot \nabla \mathbf{F} = \nabla \left(\nabla \cdot \mathbf{F} \right) - \nabla \times \left(\nabla \times \mathbf{F} \right), \tag{E.40}$$

which is valid in a curvilinear orthogonal system, including the Cartesian system as a special case. In view of this analysis, there should be no further misunderstanding about the meaning of $\nabla \cdot \nabla \mathbf{F}$ in a curvilinear orthogonal system [E6]. If one accepts our newly suggested notations for the divergence and the curl, (E.40) can be presented in a rather compact form, namely,

$$\nabla\kern-0.62em\cdot\kern0.3em \nabla \mathbf{F} = \nabla \nabla\kern-0.62em\cdot\kern0.3em \mathbf{F} - \nabla\kern-0.7em\times\kern0.1em \nabla\kern-0.7em\times\kern0.1em \mathbf{F}. \tag{E.41}$$

In conclusion, we have clarified several misunderstandings in vector analysis which have been in existence for a long time without a critical examination. It is hoped that the analysis given in this paper will facilitate the teaching of vector analysis in the future. It should be mentioned that this work was motivated by a recent study of vector analysis based on a symbolic vector method. This new work will appear elsewhere.

REFERENCES

[E1] M. Javid and P.M. Brown, *Field Analysis and Electromagnetics*. New York: McGraw-Hill, 1963.

[E2] R. Gans, *Vector Analysis* (English transl. of the 6th German ed. by W.M. Deans). London, England: Blackie & Son, 1932, p. 47.

[E3] J.W. Gibbs, "Elements of vector analysis" (privately printed, New Haven, CT, 1881, pp. 17–50; 1884, pp. 50–90), in *The Scientific Papers of J. Willard Gibbs, Vol. 3*. New York: Dover, 1961, p. 31.

[E4] W.v. Ignatowsky, *Die Vektoranalysis*. Leipzig, Germany: Dritten Auflage, Teubner, 1925.

[E5] P. Moon and D.E. Spencer, *Vectors*. Princeton, NJ: Van Nostrand, 1965.

[E6] ——, "The meaning of the vector Laplacian," *J. Franklin Inst.*, vol. 256, pp. 551–558, 1955.

[E7] P.M. Morse and H. Feshbach, *Methods of Theoretical Physics*. New York: McGraw-Hill, 1953.

[E8] H.B. Phillips, *Vector Analysis*. New York: Wiley, 1933, pp. 87–90.

[E9] C.T. Tai, "Unified definition for divergence, rotation, and gradient," *Appl. Math. Mech.* (Beijing, China), vol. 7, no. 1, pp. 1–6, 1986.

[E10] ——, "Another matter of history," *IEEE Antennas Propagat. Mag.*, vol. 33, no. 1, pp. 21–26, 1991.

REFERENCES

[1] J. Van Bladel, *Electromagnetic Fields*. New York: McGraw-Hill, 1941, Appendix 2.

[2] S.M. Candel and T. J. Poinsot, "Flame stretch and the balance equation for the flame area," *Combust. Sci. Technol.*, vol. 70, pp. 1–15, 1990.

[3] N.H. Fang and T.Y. Zhu, "Unified representation of vector functions and dyadic functions in field theory in orthogonal systems" (in Chinese), *Acta Antennie Sinica* (Beijing, China), vol. 3, no. 1, 1987.

[4] R. Gans, *Vector Analysis* (transl. from 6th German ed. by W. M. Deans). London, England: Blackie & Son, 1932.

[5] J.W. Gibbs, "*Elements of vector analysis*" (privately printed, New Haven, CT, 1881, pp. 17–50; 1884, pp. 50–90), in *The Scientific Papers of J. Willard Gibbs, Vol. 3*. New York: Dover, 1961, p. 31.

[6] E. Hallén, *Electromagnetic Theory* (transl. from Swedish ed. by R. Gaström). New York: Wiley, 1962, p. 36.

[7] H. von Helmholtz, "Das Princip der Kleinsten Wirkung in der Elektrodynamik," *Ann. Phys. U. Chem.*, vol. 47, pp. 1–26, 1892.

[8] H.A. Lorentz, *Encyklopädie der Mathematishen Wissenschaften, Band V*. Berlin, Germany: Verlag and Druck von B.G. Teubuer, 1904, part 2, p. 75.

[9] J.C. Maxwell, "A dynamical theory of electromagnetic field," in *The Science Papers of James Clerk Maxwell, Vol. 1*. Cambridge, England: Cambridge Univ. Press, 1890.

[10] P.M. Morse and H. Feshbach, *Methods of Theoretical Physics, Vol. I*. New York: McGraw-Hill, 1953.

[11] H.B. Phillips, *Vector Analysis*. New York: Wiley, 1933, p. 181.

[12] O. Reynolds, "The general equations of motion of any entity," in *Scientific Papers, Vol. III*. Cambridge, England: Cambridge Univ. Press, 1903, pp. 9–13.

[13] A. Sommerfeld, *Mechanics of Deformable Bodies*. New York: Academic, 1950, p. 132.

[14] C.T. Tai, *Dyadic Green's Functions in Electromagnetic Theory*. Scranton, PA: International Textbook, 1971.

[15] ——, "On the presentation of Maxwell's theory," *Proc. IEEE*, vol. 60, pp. 936–945, Aug. 1972.

[16] ——, "Unified definition for divergence, rotation, and gradient," *Appl. Math. Mech.*, vol. 7, no. 1, pp. 1–6, 1986.

[17] ——, "Some essential formulas in dyadic analysis," *Radio Sci.*, vol. 22, no. 7, pp. 1283–1288, 1987.

[18] C.T. Tai and N.H. Fang, "A systematic treatment of vector analysis," *IEEE Trans. Educ.*, vol. 34, no. 2, pp. 167–174, 1991.

[19] C. Truesdell and R. Toupin, "The classical field theories," in *Encyclopedia of Physics, Vol. III*. Berlin, Germany: Springer-Verlag, 1962, no. 1, p. 346.

[20] C.E. Weatherburn, *Differential Geometry*. Cambridge, England: Cambridge Univ. Press, 1927, Ch. XII.

[21] E.B. Wilson, *Vector Analysis*. New York: Charles Scribner's Sons, 1901.

INDEX

About the Author

Chen-To Tai (S'44–A'48–SM'51–F'62–LF'81) was born on December 30, 1915 in Soochow, China. He became a naturalized citizen of the United States in 1955. He obtained the B.S. degree in physics from Tsing Hua University, China, in 1937, and the D.Sc. degree in communication engineering from Harvard University in 1947.

He was a Research Fellow at Harvard from 1947–1949. He was a Senior Research Scientist at The Stanford Research Institute from 1949–1954, and an Associate Professor and later a Professor at The Ohio State University, 1954–1956 and 1960–1964. He was a Professor of Electronics at The Technical Institute of Aeronautics, Brazil, from 1956–1960. He joined The University of Michigan as a Professor in 1964, and retired from there in 1986. He is now a Professor Emeritus at The University of Michigan. During his long teaching career he served as a Visiting Professor at many universities and institutions throughout the world.

Prof. Tai is a member of the U.S. URSI Commission B and a member of the U.S. National Academy of Engineering. He was President of the IEEE Antennas and Propagation Society in 1971. At The University of Michigan, he received the EKN Outstanding Faculty Award from the Department of Electrical Engineering and Computer Science in 1971 and 1977; the Tau-Beta-Pi Outstanding Faculty Award from the College of Engineering, 1974; and the Distinguished Achievement Award from the University, 1975. He is also the recipient of the IEEE Centennial Award, 1984, and the Distinguished Achievement Award from the Antennas and Propagation Society in 1986.